Smart Cities and Japan's Energy Transition

This book offers a complex and problem-based analysis of the past, present, and future of smart cities in Japan's energy transition.

With 92% of Japanese living in urban areas and a goal of achieving net-zero greenhouse gas emissions by 2050, Japan's energy future will depend largely on how its cities can become smarter, greener, and more resilient. To reach these ambitions, a collective effort is required, with actions coming from Tokyo to Kumamoto, from Yokohama to Sapporo, and throughout dozens of smaller and bigger Japanese urban structures. This book addresses the key issues that have emerged or may emerge in various Japanese cities that are pursuing smart energy initiatives. The authors examine several issues, including international cooperation, heating decarbonisation, foreign direct investments, city planning, housing policies, and technology-related risks in the context of Japan's energy transition.

Drawing on case studies from different regions of Japan and sectors of Japanese economy significant for reaching carbon neutrality, this book will be a valuable resource for all interested in energy transition, climate action, and smart cities, where Japan and Japanese smart cities serve as excellent benchmarks.

Maciej M. Sokołowski, PhD, DSc, is a Specially Appointed Associate Professor at the Faculty of Policy Management of Keio University and is also affiliated with the Faculty of Law and Administration at the University of Warsaw. Professor Sokołowski has extensive experience in energy law and the energy sector; he has authored 100 papers and reports, including three solo books on energy regulation, combined heat and power, and the energy transition. Professor Sokołowski is a fellow of several institutions and networks, including the Sustainability College Bruges, the SI Network for Future Global Leaders, the Polish Electricity Association, the Australian Network for Japanese Law, the Japan Association of EU Studies, and the Japan Society of Public Utility Economics. Professor Sokołowski is also a lead author of the Intergovernmental Panel on Climate Change (IPCC) Special Report on Climate Change and Cities and is responsible for Chapter 4: "How to Facilitate and Accelerate Change". He has been awarded numerous distinctions, including the Swiss

Government Excellence Scholarship, the Swedish Institute Visby Programme scholarship, and the Prime Minister of Poland's Research Award. In 2024, Professor Sokołowski was named one of Stanford University's "World's Top 2% Scientists".

Fumio Shimpo, PhD, is a Professor of Law at the Faculty of Policy Management of Keio University. Professor Shimpo is an active scholar in the fields of data protection, privacy, information law, AI, and robot law in Japan. He serves as the chairperson of the Board of Directors of the Association of Law and Information Systems, the executive director of the Japanese Constitutional Law Society, a board member of the Japan Society of Information and Communication Research, the director of the Law and Computer Society, and a senior research fellow at the Institute for Information and Communications Policy of the Ministry of Internal Affairs and Communications. He was previously the commissioner for International Academic Exchange at the Personal Information Protection Commission of Japan (2018–2023) and the former vice-chair of the OECD Working Party on Security and Privacy in the Digital Economy (2009–2016).

Routledge Explorations in Energy Studies

Northern Indigenous Community-Led Disaster Management and Sustainable Energy
Ranjan Datta, Margot Hurlbert and William Marion

Energy Policy Design in the South-Eastern Mediterranean Basin
A Roadmap to Energy Efficiency
Bertug Ozarisoy and Hasim Altan

Living with Energy Poverty
Perspectives from the Global North and South
Edited by Paola Velasco-Herrejón, Breffní Lennon and Niall Dunphy

Transitioning Fossil-Based Economies
Sustainable Strategies for Energy Change
Hassan Qudrat-Ullah

The Future of Liquified Natural Gas in a Decarbonising World
Omran Al-Kuwari

Energy Use in Bitcoin Mining
The Environmental Impact of Cryptocurrencies
Benjamin A. Jones; Andrew Goodkind and Robert P. Berrens

Smart Cities and Japan's Energy Transition
Past, Present, and Future
Edited by Maciej M. Sokołowski and Fumio Shimpo

The Energy Transition in Japan
Smart Cities and Smart Solutions
Edited by Maciej M. Sokołowski and Fumio Shimpo

For more information about this series, please visit: www.routledge.com/Routledge-Explorations-in-Energy-Studies/book-series/REENS

Smart Cities and Japan's Energy Transition

Past, Present, and Future

Edited by Maciej M. Sokołowski and Fumio Shimpo

LONDON AND NEW YORK

First published 2025
by Routledge
4 Park Square, Milton Park, Abingdon, Oxon OX14 4RN

and by Routledge
605 Third Avenue, New York, NY 10158

Routledge is an imprint of the Taylor & Francis Group, an informa business

© 2025 selection and editorial matter, Maciej M. Sokołowski and Fumio Shimpo; individual chapters, the contributors

The right of Maciej M. Sokołowski and Fumio Shimpo to be identified as the authors of the editorial material, and of the authors for their individual chapters, has been asserted in accordance with sections 77 and 78 of the Copyright, Designs and Patents Act 1988.

All rights reserved. No part of this book may be reprinted or reproduced or utilised in any form or by any electronic, mechanical, or other means, now known or hereafter invented, including photocopying and recording, or in any information storage or retrieval system, without permission in writing from the publishers.

Trademark notice: Product or corporate names may be trademarks or registered trademarks, and are used only for identification and explanation without intent to infringe.

British Library Cataloguing-in-Publication Data
A catalogue record for this book is available from the British Library

ISBN: 978-1-032-74900-6 (hbk)
ISBN: 978-1-032-74901-3 (pbk)
ISBN: 978-1-003-47144-8 (ebk)

DOI: 10.4324/9781003471448

Typeset in Optima
by Apex CoVantage, LLC

Contents

Contributors

Muneki Adachi is a senior negotiator for climate change in the Climate Change Division, International Cooperation Bureau, Ministry of Foreign Affairs, Japan. He has engaged in international climate negotiations at COP 26–29 and relevant UN conferences, including his role as the lead negotiator of the Government of Japan on the 1st global stocktake at COP28. His research focuses on the linkage between climate science and international politics.

Yuki Akiyama is a professor of spatial information science and urban studies at Tokyo City University, Faculty of Architecture and Urban Design. He holds a PhD in environmental studies from the University of Tokyo. Professor Akiyama has published extensively in peer-reviewed journals, focusing on spatial information, urban planning, and data science. His research encompasses areas such as urban transportation management, geography, real estate science, and micro geodata analysis for disaster prevention. Professor Akiyama has received numerous awards from domestic and international academic societies. He is also a frequent speaker at international conferences.

Damian M. Bielicki, PhD, is a senior lecturer in law and the director of the Law, Technology and Society Research Group at Kingston University London, UK. Damian specialises in law and technology. In addition to his post at Kingston University, he is also a lecturer in space law and cyber law at Birkbeck University of London, UK. He is a senior fellow of the Higher Education Academy, a member of the International Law Association, and a member of the International Institute of Space Law. He is also the founder of the Space Law Centre (https://www.spacelawcentre.org/), the world's largest online database of space law and policy-related issues.

Jordan Carlson, PhD, is a postdoctoral researcher in energy geography. He holds a doctorate from Queen's University (Kingston, Ontario, Canada) and is a research associate at the Center for Energy Systems Design, International Institute for Carbon Neutral Energy Research, Kyushu University (Fukuoka, Japan). Dr Carlson has published on renewable energy development policy as well as decarbonisation challenges arising from different hydrogen production and use methods. His research focuses on understanding the ways that existing and emerging technologies can contribute to diverse energy transitions in different locations around the world. Prior to joining Kyushu University, he was awarded a two-year joint Japan Society for the Promotion of Science and Natural Sciences and Engineering Research Council of Canada International Postdoctoral Fellowship to study decarbonisation risks and hydrogen at Kyoto University.

Jesper Edman, PhD, is a professor of management at Waseda University. His research focuses on the internationalisation strategies of Japanese companies and industries. Professor Edman is an author of numerous publications on international business management and a frequent speaker on internationalisation strategies. He also consults for leading Japanese and foreign enterprises as well as government entities.

Fukuzo Hasegawa is an associate professor of administrative law and public law teaching full time at the Faculty of Policy Management, Keio University. He holds Juris Doctor and Master of Laws degrees. He is a licensed attorney in Japan (Tokyo Bar Association). He has co-authored a systematic book on administrative law and published articles on administrative planning and the digitisation of public administration. He is also engaged in research on the institutional design and legal reform of German administrative law, since the theoretical structure of Japanese administrative law is organised on the basis of the German system.

Carin Holroyd is a professor of political studies at the University of Saskatchewan (Canada). She holds a PhD in political science from the University of Waikato (New Zealand) and an MSc in Japanese business studies from Chaminade University of Honolulu (US)/Sophia University (Japan). She has been teaching in the Department of Political Studies at the University of Saskatchewan since 2012 and was previously in the Department of Political Science at the University of Waterloo (Canada). Carin has published extensively, particularly in the field of Japanese political economy, including eight scholarly books and more than 40 articles.

Hiroshi Ito, PhD, is a professor of international and sustainable development and education. He teaches at Nagoya University of Commerce and Business (NUCB). Before joining NUCB, he served as an education expert with UNESCO (France), UNICEF (Philippines), and JICA (Ecuador and Paraguay). His research interests lie in educational assessment and environmental policy, and he has several publications in these fields.

Klaudia Pryjmak is a researcher of law, holding a degree in sociology, and is associated with the Faculty of Law and Administration at the University of Warsaw. Her research focuses on tax and international trade law, particularly the green taxes and the contribution of the European Union's tax policy to achieving the climate goals. She is also an expert on the EU CBAM mechanism, on which she advises professionally and authors academic publications.

Kie Sanada is an assistant professor of Culture, Society, and Media at Ritsumeikan Asia Pacific University, College of Asia Pacific Studies. She also holds a position as an affiliated researcher at the European Institute of Japanese Studies, Stockholm School of Economics. She received her PhD in global and area studies from Humboldt University Berlin. Her research focuses on elucidating the mechanisms through which projects of social innovation lead to wider regional development. Her recent publications extensively address Japanese smart city projects.

Fumio Shimpo, PhD, is a professor of law at the Faculty of Policy Management of Keio University. Professor Shimpo is an active scholar in the fields of data protection, privacy, information law, AI, and robot law in Japan. He serves as the chairperson of the Board of Directors of the Association of Law and Information Systems, the executive director of the Japanese Constitutional Law Society, a board member of the Japan Society of Information and Communication Research, the director of the Law and Computer Society, and a senior research fellow at the Institute for Information and Communications Policy of the Ministry of Internal Affairs and Communications. He was previously the commissioner for International Academic Exchange at the Personal Information Protection Commission of Japan (2018–2023) and the former vice-chair of the OECD Working Party on Security and Privacy in the Digital Economy (2009–2016).

Maciej M. Sokołowski, PhD, DSc, is a specially appointed associate professor at the Faculty of Policy Management of Keio University, also affiliated with the Faculty of Law and Administration at the University of Warsaw. Professor Sokołowski has extensive experience in energy law and the energy sector; he has authored 100 papers and reports, including three solo books on energy regulation, combined heat and power, and the energy transition. Professor Sokołowski is a fellow of several institutions and networks, including the Sustainability College Bruges, the SI Network for Future Global Leaders, the Polish Electricity Association, the Australian Network for Japanese Law, the Japan Association of EU Studies, and the Japan Society of Public Utility Economics. Professor Sokołowski is also a lead author of the Intergovernmental Panel on Climate Change (IPCC) Special Report on Climate Change and Cities and is responsible for Chapter 4: "How to Facilitate and Accelerate Change". He has been awarded numerous distinctions, including the Swiss Government Excellence Scholarship, the Swedish Institute Visby Programme scholarship, and the Prime Minister of Poland's Research Award. In 2024, Professor Sokołowski was named one of Stanford University's "World's Top 2% Scientists".

Makoto Tajima, director/senior research fellow of the Institute for Sustainable Energy Policies (ISEP), holds an MSc in agronomy from the University of Hawaii, specialising in natural resource management. With over 25 years of experience in international development, he has worked for NGOs and the Japan International Cooperation Agency (JICA). He advises on Disaster Risk Reduction and Response at the Japan NGO Center for International Cooperation (JANIC). He is also on the Japan Solar Sharing Federation (JSSF), CWS Japan, and Women's Eye boards. His focus is on international R&D collaboration in agrivoltaics and district heating. He has been on the Scientific Committee of the AgriVoltaics International Conference since 2020 and on the Principal Action Committee of the India Agrivoltaics Alliance (IAA) since 2023.

Gregory Trencher is an associate professor of decarbonisation and sustainability governance. He holds a PhD in sustainability science and conducts research at Kyoto University, Graduate School of Global Environmental Studies. He currently serves on the editorial board for *Energy Research & Social Science* and as an editor for *Frontiers in Sustainable Cities*. He has published over 50 peer-reviewed papers in journals such as *Nature Communications, One Earth, Environmental Research Letters, Energy Policy, Energy Research and Social Science, PLOS One, Climatic Change,* and *Technological Forecasting and Social Change*. His research and opinion have appeared in media outlets such as *The Guardian, New York Times, Nikkei Asia, Channel NewsAsia (CNA), ABC Australia,* and *Carbon Brief*.

Yuichiro Tsuji is a professor of law at the Graduate School of Law and Faculty of Law at Meiji University. His research focuses on constitutional law, administrative law, comparative law, and climate change law. He holds an LLM and a JSD (Doctor of the Science of Jurisprudence) from UC Berkeley Law. He completed his Master of Law from Kyoto University Graduate School of Law. He has published three monographs in Japanese and several books as editor for US environmental law and climate change law.

Marco Zappa is an assistant professor of Japanese studies at Ca' Foscari University of Venice, Italy. A Japanese specialist by training, he has conducted research on forms of international cooperation in Asia. Specifically, his research focuses on human resource development, knowledge transfer and policy circulation in the urban realm in development initiatives across East and Southeast Asia. He has published several articles and volume chapters on the history of international relations and development in Northeast and Southeast Asia.

Foreword

It is with great pleasure and respect that I write this foreword for a book judiciously edited and compiled by Dr Maciej M. Sokołowski and Dr Fumio Shimpo. Maciej and Fumio have the good foresight to develop a tome regarding an important topic in urban climate change and sustainability – on smart cities in Japan and how they play a significant role in enabling Japan's energy transition. It is also laudable that Maciej and Fumio have managed to get a very strong list of academic contributors to address one of the urgent challenges of our time.

Smart cities represent a crucial frontier in the pursuit of sustainability, and Japan stands at the forefront of this movement. The country's innovative approaches and successful implementations of smart city projects have made it a global leader, with cities like Kumamoto, Shin-Sapporo, Toyama, and Yokohama serving as models of urban efficiency, sustainability, and technological integration towards the energy transition. Japan's achievements in this area are the envy of many, and the themes and cases presented in the chapters of this book will undoubtedly offer invaluable insights.

At the heart of this book lies the recognition that the energy transition is not merely a technical shift but a pivotal component for both climate action and the implementation of the Sustainable Development Goals in Japan's urban areas. The urgent transition needed to develop more sustainable energy systems or better urban planning and designs that reduce emissions and increase the efficiency of resource use cannot be overstated. These factors are critical in mitigating climate change and ensuring a resilient future for generations of urban residents to come in Japan. The strategies to enable action and frameworks to reduce barriers outlined in this book will serve as a catalyst, helping to build the momentum necessary to drive change on a larger scale.

Maciej and Fumio have masterfully curated a collection of chapters that not only show Japan's leadership in smart cities but also provide a roadmap for others to follow. The interdisciplinary nature of this work, touching on planning, finance, legal, and other related topics, makes it accessible to a broad audience. Readers, whether they are scientists, policymakers, or urban planners, will find in these pages a wealth of knowledge on what works and what can be replicated or adapted in other contexts. It bridges the gap between theoretical research and practical application. I have no doubt this book is a valuable resource for anyone committed to understanding and advancing the role of smart cities in the energy transition.

In closing, I would like to extend my congratulations to Maciej and Fumio and their team of authors for this remarkable achievement. Their dedication and vision for a sustainable future are evident in this book's pages. I am confident that this work will inform and hopefully inspire the next generation of scholars and practitioners who will continue to push the boundaries of what is possible in smart cities and energy systems.

Winston T.L. Chow, Ph.D.
Professor of Urban Climate & Co-Chair,
Intergovernmental Panel on Climate Change
Working Group II

Preface

With 92% of Japanese living in urban areas and a goal of achieving net-zero greenhouse gas emissions by 2050, Japan's energy future will depend largely on how its cities can become smarter, greener, and more resilient. To reach these ambitions, all hands on deck are needed, with actions coming from Tokyo to Kumamoto, from Yokohama to Sapporo, and throughout dozens of Japanese smaller and larger urban structures. Conducting the energy transition, however, is not an easy task and over the years, this process has faced numerous difficulties.

In this light, this book offers a comprehensive and problem-based analysis of the past, present, and future of smart cities in Japan's energy transition. It examines the key issues that have emerged or may emerge in various Japanese cities that are pursuing smart energy initiatives. This relates to such matters as the Sustainable Development Goals, local decarbonisation, foreign direct investment, international cooperation, urban planning, or local policies and regulatory frameworks.

Drawing on case studies from different regions and fields significant for reaching carbon neutrality, this book provides a unique and holistic perspective on the challenges of smart cities in Japan's energy transition. It is a valuable resource for anyone interested in learning more about the current state and prospects of smart cities in Japan and their role in energy transformation.

Maciej M. Sokołowski
Fumio Shimpo

Acknowledgements

Hereby, we extend our sincere gratitude to all authors who contributed to this book project. We appreciate your invaluable contributions and look forward to future collaborations. We are very grateful for the kind words of Dr Winston Chow, Co-Chair of the Intergovernmental Panel on Climate Change Working Group II, who kindly agreed to write the Foreword for our book.

We thank Ms Klaudia Pryjmak for serving as assistant editor and the initial proofreader of the manuscript. We also thank Ms Justyna Kamila Kanas for providing useful reference materials essential to the research foundation of this project.

Special thanks go to all colleagues from the JST Moonshot R&D project, JPMJMS2215, including Ms Kimie Hatakeyama. Your invaluable support and assistance have been extremely valuable to us.

Last but not least, we thank all the editors at Routledge who were involved in publishing this book. As always, you did an excellent job.

本書が多くの方々の支えと協力によって完成したことに改めて感謝申し上げます。この場を借りて、心より深い感謝の意を表します。

Maciej M. Sokołowski
Fumio Shimpo

1 Smart Cities and Japan's Energy Transition

Connecting Yesterday with Tomorrow

Maciej M. Sokołowski and Fumio Shimpo

1. Introduction

The global discourse on Japan is often twofold, driven by two opposing narratives. On the one hand, some claim that it is the country of the past, a former number two economy that has never reached first place – a fallen economy hit by the 1990s bubble from which it has never recovered. This perspective stems from Japan's economic stagnation post the 1990s bubble, often summarised by terms such as the Lost Decade, followed by social issues like depopulation or natural disasters like the catastrophic events of March 2011, and other problems including recent events like the postponed to 2021 Tokyo Olympics or the growing inflation of 2022 and 2023.

On the contrary, others strongly believe that Japan remains the country of the future – the land of tomorrow where the sun rises and where every city resembles Tokyo. This Japan is characterised by its rapid adoption of cutting-edge technology, featuring urban interactive screens on streets teeming with autonomous vehicles, bullet trains streaking through the Japanese metropolises, and robotics not only integrated into industrial processes but also seamlessly embedded within everyday life – a society where technology and human interaction are flawlessly connected.

The truth, as in many instances, lies somewhere in-between. While Japan aims to implement new technologies and create the cities of tomorrow, it struggles with a variety of issues ranging from an ageing society to bureaucratisation. The same reflection is true for the country's efforts to conduct the energy transition, where the deployment of green technologies contrasts with new coal-fired plants entering operation.

2. Book's Structure and Issues Discussed

The book is structured into 13 chapters. Skipping this introductory part, Chapter 2, contributed by Muneki Adachi and Klaudia Pryjmak, explores the significance of cities, policies, and practices in global efforts to tackle climate change and reduce greenhouse gas (GHG) emissions under frameworks such as the City-to-City Collaboration Programme and the Clean City Partnership

DOI: 10.4324/9781003471448-1

Programme. In this context, this chapter examines how communities in Japan are engaging in sustainability through various policy frameworks and practical actions for transition to cleaner energy sources. In addition, the chapter evaluates both current and prospective actions designed to address climate challenges, providing an analysis of their impact and scope for expansion.

In Chapter 3, Jesper Edman discusses how foreign direct investment (FDI) can contribute to the development of sustainable smart cities in Japan. He highlights how inward FDI plays a critical role in transferring capital, technology, and knowledge. In addition, he shows how outward FDI offers a means to enhance learning and develop new capabilities, which can subsequently be reapplied within Japan. Moreover, this chapter addresses how the combination of inward and outward FDI fosters significant network linkages between local smart cities in Japan and similar municipalities abroad. As suggested in the chapter, these network linkages have the potential to boost the dynamism and innovative capacity of smart cities, while simultaneously promoting knowledge transfer and development.

Chapter 4, authored by Jordan T. Carlson and Gregory Trencher, addresses the risks associated with decarbonisation objectives due to reliance on sophisticated, costly, and unproven technologies. In this light, the chapter examines the technological approaches Japan has adopted in pursuit of decarbonisation within its smart cities to date. By focusing on three areas that Japan has emphasised in its smart cities – mobility, energy management, and the built environment – the chapter discusses whether alternative approaches could provide equal or greater decarbonisation benefits.

In Chapter 5, Kie Sanada and Marco Zappa survey smart city projects in Japan employing Kanemoto and Tokuoka's Urban Employment Area model to highlight persisting inequalities and imbalances between urban areas and regions. Moreover, this chapter shows the achievement of the second stage of smart innovation in terms of structural and administrative change towards the materialisation of Society 5.0. In doing so, this chapter provides a multi-level perspective on the decentralisation process, clarifying the role of various actors involved in a complex sociotechnical network like smart cities, which serve as a good example of structural transformations in Japan.

In Chapter 6, Yuki Akiyama introduces a methodology for estimating the distribution of vacant houses using municipal data. The research findings aim to aid in mapping the spatial distribution of vacant houses across extensive areas – an issue that can contribute to the realisation of smart cities where vacant houses can be transformed into energy-efficient homes through renovations. Moreover, the municipal data used in this chapter are anticipated to provide a universally applicable method for all Japanese municipalities, considering the nationwide maintenance of similar data.

Chapter 7, written by Fukuzo Hasegawa, reviews the Japanese policy agenda on smart cities and examines legal issues pertaining to the role of smart cities in the energy transition towards carbon neutrality in Japan. Furthermore, given the benefits of referencing international examples, this chapter offers an analysis of energy transition and urban management in Germany, offering solutions to challenges faced by Japanese smart cities. These include areas such as public participation and local cooperation or public management with integration of planning.

Chapter 8, provided by Makoto Tajima, also refers to international benchmarks for Japan by comparing the Danish district heating with the Japanese heating sector. In this context, the chapter addresses heat decarbonisation in Japan as a solution for future smart compact cities. The chapter proposes a potential model for a smart compact city, referring to Ogata as a case study for further extension into typical urban structures. This chapter shows the benefits of this development, contrasted with a range of constraints that may block this type of project.

In Chapter 9, Damian M. Bielicki examines the application of data gathered from space-based satellites to the energy sector. While numerous case studies from around the world are provided to support his analysis, a particular focus is given to Japan and its Basic Plan on Space Policy. These examples identify the key advantages of using space technologies in the energy sector, and also discuss the challenges associated with the operation of satellites in space.

Chapter 10, offered by Hiroshi Ito, examines Japan's approaches and initiatives towards smart cities, along with an exploration of how these align with the Sustainable Development Goals (SDGs). This chapter presents a comprehensive analysis of the strategies employed by Japan's three cities: Toyama, Sapporo, and Kumamoto. This chapter contrasts and compares the shared and distinctive aspects of their commitment to sustainability, innovation, and resilience emphasising the global implications for smart city development and highlighting the role of these cities as exemplars of sustainable urban planning.

In Chapter 11, Carin Holroyd focuses on one of the smart cities in Japan: Sapporo. She highlights the Shin Sapporo Smart City project in terms of energy savings and disaster resilience together with its reliance on a natural gas cogeneration system and an artificial intelligence (AI)-driven Community Energy Management System. This chapter also provides the lessons that can be learned from this new smart city project. This especially concerns Sapporo's experience with addressing energy needs during severe winters, a reality that the city shares with northern communities around the world.

Chapter 12, authored by Yuichiro Tsuji, examines Yokohama's smart city model from a constitutional and administrative law perspective. This chapter analyses Yokohama's goals for GHG reduction and plans related to low-carbon technologies used in energy, buildings, and transportation under the framework of a smart city. This chapter presents this from the perspective of the active participation of the private sector required to promote smart and climate-friendly urban solutions, discussing legal problems related to democratic legitimacy that reflect the voices of Yokohama's citizens in the design of the smart city.

Finally, in Chapter 13, the editors of this book discuss the capital of Japan – Tokyo within the context of its smart city policies, focusing on sustainability. This chapter explores Tokyo's historical environmental issues, such as pollution and contamination during the 1960s and 1970s. This chapter also reviews Tokyo's strategies, programmes, and projects to identify those with the potential to enhance the sustainability and innovation of Japan's capital. Lastly, the chapter addresses the relationship between being a smart city and a sustainable city, using Tokyo as an example.

3. Key Points for Deeper Insight

Japan was one of the first countries engaged in sustainable city movement, with initial discourse within the Japanese administration regarding the feasibility of an environmentally friendly city dating back to the late 1980s (Holroyd, 2025). The actions to improve urban environmental conditions related to air or water quality were, however, implemented much earlier, in the 1960s and 1970s (Sokołowski and Shimpo, 2025). For instance, in 1978, the Government of Japan established the Garden City State Initiative aimed at improving residents' quality of life by bringing rural harmony into urban structures and creating a *garden city* (Hasegawa, 2025). Earlier in the 1970s, the term *compact city* was introduced to Japan; however, deeper research on it was not conducted until the 1990s (Tajima, 2025). The early 1990s were also a moment, when the term *smart city* emerged (Ito, 2025). In 1997, the Japanese government launched the Eco-Town Project, which began designating Japan's first officially recognised eco-towns, although the term itself had appeared earlier (see Low, 2013, pp. 12–15).

The range of frameworks demonstrates a prioritisation of different aspects, distinguished by their names. Within this spectrum, the smart city concept, however, offers a much-needed breadth and depth, enabling easy adaptation. Moreover, many smart cities have an energy-related focus (like *smart energy city*), prioritising energy efficiency, lighting systems, zero-emission buildings, or low-carbon transportation, enhanced by AI-driven solutions, Internet of Things (IoT), sensor technologies, and smart metering within city infrastructure (see Ito, 2025). The significance of these areas increases in tandem with the expansion of cities globally, alongside their impact on the climate and, conversely, the adverse effects of climate change on urban residents, particularly through intensified heat extremes (see Adachi and Pryjmak, 2025).

While smart cities address a variety of increasingly complex issues, the question of energy and sustainability in urban areas, in particular, is multi-scalar, involving various factors and drivers (see Sokołowski and Visvizi, 2023). As showed in Japan, for instance, within the concept of Digital Garden City Nation, or under the plans and strategies adopted by the Japanese capital, or within the idea of Society 5.0, these fields (energy and sustainability) are linked to the environment-climate-energy-technology axis. This also highlights the close positions of green transformation and digital transformation and various interrelations between these two platforms (see Sanada and Zappa, 2025; Sokołowski and Shimpo, 2025; Akiyama, 2025; Bielicki, 2025). Concrete projects in these fields are being implemented in Japanese cities, as exemplified by Kumamoto, Sapporo, Tokyo, Toyama, Yokohama, and many others (see Holroyd, 2025; Ito, 2025; Sokołowski and Shimpo, 2025; Tsuji, 2025). This also includes future-oriented actions like creating digital smart cities that recreate the urban functions in virtual reality, realising the concept of a digital twin environment, and offering a valuable platform for testing and introducing the concept of cybernetic avatars (see Sokołowski and Shimpo, 2025).

The wide agenda on smart cities in Japan, however, does not imply that no further action is required. As urban climate action gains momentum, further investments in research and development, fair resource allocation, and adaptive strategies are needed to achieve long-term success (Adachi and Pryjmak, 2025). The same is also true for Japan. Although Japan has significantly invested in pioneering and testing new technologies as part of its decarbonisation efforts through smart city initiatives, several challenges hinder the achievement of these decarbonisation goals (see Carlson and Trencher, 2025; Sanada and Zappa, 2025). This, for example, concerns leveraging the potential of different stakeholders to become partners in smart city development and increasing the active participation of the private sector to promote smart and climate-friendly urban solutions (Edman, 2025; Tsuji, 2025).

In addition, greater emphasis should be placed on the human aspect of Japanese smart cities. As a renowned Japanese architect, Kengo Kuma (2021) notes, "Japan tends to focus on building the framework and hardware beginning and then we address the human aspect". Tackling this matter includes ensuring effective communication with residents, not only to inform them of the local administration's intentions but also to listen to their views and feedback and adapt plans accordingly (Hasegawa, 2025). "Increased awareness and participation in the smart city planning process is a win-win option", concludes Tsuji (2025). "First, you need to establish smart citizens and that will naturally nurture a smart city", observes Kengo Kuma (2021).

Acknowledgements

This research was supported by the JST Moonshot R&D project, JPMJMS2215.

References

Adachi, M. and Pryjmak, K. (2025) "Japanese Cities in the Context of Local Decarbonisation and International Cooperation", in M.M. Sokołowski and F. Shimpo (eds) *Smart Cities and Japan's Energy Transition: Past, Present, and Future*. London: Routledge, pp. 7–30.

Akiyama, Y. (2025) "Learning the Characteristics of Vacant Houses: Smart Solutions for Japanese Municipalities", in M.M. Sokołowski and F. Shimpo (eds) *Smart Cities and Japan's Energy Transition: Past, Present, and Future*. London: Routledge, pp. 90–114.

Bielicki, D.M. (2025) "Satellite Applications for Sustainable Energy", in M.M. Sokołowski and F. Shimpo (eds) *Smart Cities and Japan's Energy Transition: Past, Present, and Future*. London: Routledge, pp. 153–164.

Carlson, J.T. and Trencher, G. (2025) "'Smart' Cities and Dumb Solutions: The Risks of Technology-Reliant Solutions for Decarbonisation in Japan", in M.M. Sokołowski and F. Shimpo (eds) *Smart Cities and Japan's Energy Transition: Past, Present, and Future*. London: Routledge, pp. 48–70.

Edman, J. (2025) "The Role of Inward and Outward FDI in Sustainable Smart Cities", in M.M. Sokołowski and F. Shimpo (eds) *Smart Cities and Japan's Energy Transition: Past, Present, and Future*. London: Routledge, pp. 31–47.

Hasegawa, F. (2025) "Legal System and Public Policy of Smart Cities in Energy Transition: Germany: Japan Contexts", in M.M. Sokołowski and F. Shimpo (eds) *Smart Cities and Japan's Energy Transition: Past, Present, and Future*. London: Routledge, pp. 115–130.

Holroyd, C. (2025) "Smart City Development Underway: Lessons From Shin Sapporo's Smart City", in M.M. Sokołowski and F. Shimpo (eds) *Smart Cities and Japan's Energy Transition: Past, Present, and Future*. London: Routledge, pp. 188–201.

Ito, H. (2025) "Energy Transition in Japan's SDGs Future Cities: Toyama, Sapporo and Kumamoto", in M.M. Sokołowski and F. Shimpo (eds) *Smart Cities and Japan's Energy Transition: Past, Present, and Future*. London: Routledge, pp. 165–187.

Kengo, K. (2021) "Exchange of Opinions: Minutes of the Tokyo eSG Strategy Board". Available at: https://www.tokyobayesg.metro.tokyo.lg.jp (Accessed: 31 August 2024).

Low, M. (2013) "Eco-Cities in Japan: Past and Future", *Journal of Urban Technology*, 20(1), pp. 7–22.

Sanada, K. and Zappa, M. (2025) "Deepening the City-Region Divide in 21st Century Japan: Smart Cities as a Tool to Achieve Administrative Neoliberalisation", in M.M. Sokołowski and F. Shimpo (eds) *Smart Cities and Japan's Energy Transition: Past, Present, and Future*. London: Routledge, pp. 71–89.

Sokołowski, M.M. and Shimpo, F. (2025) "Sustainable Smart City Tokyo: Between Problems of the Past and Chances of the Future", in M.M. Sokołowski and F. Shimpo (eds) *Smart Cities and Japan's Energy Transition: Past, Present, and Future*. London: Routledge, pp. 217–238.

Sokołowski, M.M. and Visvizi, A. (2023) "Exploring the Smart Cities: Energy Communities Nexus", in M.M. Sokołowski and A. Visvizi (eds) *Routledge Handbook of Energy Communities and Smart Cities*. London: Routledge, pp. 1–10.

Tajima, M. (2025) "Heat Decarbonisation: A Solution for the Future Smart Compact Cities in Japan", in M.M. Sokołowski and F. Shimpo (eds) *Smart Cities and Japan's Energy Transition: Past, Present, and Future*. London: Routledge, pp. 131–152.

Tsuji, Y. (2025) "Democratic Legitimacy of the Smart City: A Case Study of Yokohama", in M.M. Sokołowski and F. Shimpo (eds) *Smart Cities and Japan's Energy Transition: Past, Present, and Future*. London: Routledge, pp. 202–216.

2 Japanese Smart Cities in the Context of Local Decarbonisation and International Cooperation

Muneki Adachi and Klaudia Pryjmak

1. Introduction

For decades, cities have garnered attention across disciplines – science, politics, and business – at local, national, regional, and international levels in the context of climate change. Current urban-related research, strategies, and implementation predominantly focus on climate change adaptation and mitigation. Adaptation involves adjusting to the impacts of a changing climate, encompassing human interventions and adjustments in natural systems to mitigate adverse effects and exploit potential benefits. By contrast, mitigation efforts address the root causes of climate change by reducing greenhouse gas (GHG) emissions as highlighted by the Asia-Pacific Climate Change Adaptation Information Platform (AP-PLAT, n.d.). The Intergovernmental Panel on Climate Change (IPCC) has published the Sixth Assessment Report (AR6), highlighting the important role of cities in climate change mitigation. Working Group III (WG3) of the IPCC, focused on Mitigation of Climate Change, has contributed updated guidelines on cities' roles in mitigating climate change, including the Synthesis Report (IPCC, 2023). Moreover, the IPCC has decided to prepare a special report entitled "Climate Change and Cities" in the Seventh Assessment Report (AR7). From the perspective of international politics, the United Nations Framework Convention on Climate Change (UNFCCC) has completed its first global stocktake (GST), acknowledging the significant role and active engagement of cities and other stakeholders in climate change mitigation and adaptation. In addition, multidisciplinary and multilateral declarations, such as the G7's Roundtable on Subnational Climate Actions in 2023, have underscored the importance of urban climate action.

Drawing upon these global initiatives, the Japanese government and the Ministry of the Environment of Japan (MOEJ) have been actively addressing climate change with a distinct focus on cities, encompassing both local decarbonisation efforts and international cooperation. To illustrate the pivotal role of cities, policies, and good practices in international cooperation for climate change mitigation and decarbonisation, the first part of this chapter presents relevant policy developments and actions undertaken globally. It

DOI: 10.4324/9781003471448-2

highlights key narratives on cities and climate change based on recent scientific findings and summarises political decisions made under the UNFCCC, the Paris Agreement, and other multilateral declarations. The second part of this chapter delves into the current state of decarbonisation and international cooperation within local communities in Japan. It begins by introducing the primary objectives of current policies and actions aimed at fostering decarbonisation within the country. Subsequently, it shows examples of successful actions and best practices established by Japan to leverage international cooperation for the benefit of local communities. Finally, the chapter concludes by highlighting ongoing and future initiatives to mitigate and adapt to climate change, offering insights into their effectiveness and potential for further development.

2. Cities as Key Actors in Achieving Global Net-Zero Greenhouse Gas Emissions: Emerging Narratives

This section covers the IPCC's key findings on cities, along with the global research and action agenda on cities, as well as relevant international frameworks on these issues under the Paris Agreement.

2.1 *IPCC Findings on Cities: AR6 Synthesis Report*

The IPCC WG 3 during the AR6 process emphasised that urban areas have significant potential to create opportunities that enhance resource efficiency and decrease GHG emissions (IPCC, 2022). Urban infrastructure and architecture development strategies are focused on reducing emissions in cities, with the overarching objective of reaching net-zero emissions. The AR6 findings suggest that, in cities that are highly developed, expanding quickly, or on the rise, efforts to reach net-zero emissions will include reducing energy and material consumption, transitioning to electric energy, and enhancing carbon sequestration in urban areas. Cities have the potential to reach these ambitious objectives, but only by cutting emissions both within and beyond their administrative borders through tackling the GHG emissions of the linked supply chains, which has been acknowledged to have positive ripple effects in various sectors (IPCC, 2022).

According to the IPCC (2023), global urban carbon dioxide (CO_2) and methane (CH_4) emissions linked to consumption are projected to rise significantly, from 29 $GtCO_2$-eq[1] in 2020 to 34 $GtCO_2$-eq by 2050, even under a scenario of moderate mitigation efforts. This "middle-of-the-road" scenario, referred to as SSP2-4.5 within the Shared Socioeconomic Pathways (SSP) framework of the AR6 Report (IPCC, 2021), assumes a continuation of historical trends in social, economic, and technological development. Under this scenario, the steady increase in GHG emissions would lead to a projected warming of 2.5–3°C by 2100. However, the IPCC warns that without significant mitigation actions, urban CO_2 and CH_4 emissions could escalate to 40 $GtCO_2$-eq by 2050, representing a high-emissions scenario (SSP3-7.0) with potentially more severe consequences for global warming. However, through ambitious and rapid mitigation efforts, like increased electrification and enhanced energy and material efficiency, global urban CO_2 and CH_4 emissions based on consumption could be decreased to 3 $GtCO_2$-eq by 2050 in the modelled scenario with very low GHG emissions (SSP1-1.9). The Report also points out that the sequence and effectiveness of actions to reduce GHG emissions will vary depending on a city's land use, spatial arrangement, development phase, and degree of urbanisation (IPCC, 2021).

As indicated, effective strategies for well-established cities to significantly decrease GHG emissions include improving building efficiency, repurposing or retrofitting existing structures, strategically utilising urban spaces, and encouraging non-motorised transportation like walking and cycling, as well as public transport such as buses and trams. Nevertheless, appropriate tactics are essential to meet the considerable infrastructure development demands for improving the quality of life. This can be achieved through energy-efficient infrastructures and services, along with urban design that prioritises people. The IPCC (2022) discusses three of the following strategies "to be effective when implemented concurrently":

1. reducing or changing energy and material use towards more sustainable production and consumption;
2. electrification in combination with switching to low-emission energy sources; and
3. enhancing carbon uptake and storage in the urban environment, for example, through bio-based building materials, permeable surfaces, green roofs, trees, green spaces, rivers, ponds, and lakes.

Regarding cities, the Summary for Policymakers by IPCC's WG3 (2022) further reveals that the application of diverse mitigation strategies at the urban level can have far-reaching effects on different sectors, causing a decrease in GHG emissions within and beyond a city's administrative limits. As highlighted by the WG3, cities' ability to create and apply mitigation strategies differs based on overall regulations, institutional frameworks, and facilitating factors such as access to financial and technological resources, local governance capabilities, involvement of civil society, and municipal budget authority. When it comes to achieving net-zero GHG emissions at the city level, the report highlights that an increasing number of cities are establishing climate goals, such as net-zero GHG targets. Given the regional and global reach of urban consumption patterns and supply chains, the full potential for reducing consumption-based urban emissions to net zero can be met only when emissions beyond cities' administrative boundaries are also addressed in these goals and targets (IPCC, 2022).

The efficiency of these methods is contingent on cooperation and coordination with national and sub-national governments, industry, and civil society, and on whether cities have the appropriate capacity to plan and implement mitigation strategies. Urban areas can play a significant role in decreasing emissions within supply chains that extend past their city limits. One way to achieve this is by implementing building codes and selecting sustainable construction materials.

The IPCC's AR6 Synthesis Report (2023) emphasises the crucial role of cities, echoing the findings from WG3, by examining current urban approaches to climate adaptation and mitigation aimed at addressing the detrimental effects of climate change on residents. In urban areas, observed climate change has already caused adverse impacts on human health, livelihoods, and key infrastructure, particularly through intensified hot extremes. Critical systems such as transportation, water, sanitation, and energy have been negatively affected by both acute extreme weather events and insidious slow-onset processes, resulting in significant economic losses, service disruptions, and adverse effects on overall well-being, thereby disproportionately impacting marginalised populations.

As for addressing these challenges, the AR6 Report (IPCC, 2023) highlights the effectiveness of ecosystem-based adaptation approaches, such as urban greening initiatives, wetland restoration, and upstream forest ecosystem conservation and rehabilitation, in mitigating flood risks and urban heat island effects (high confidence). The Report also notes the moderate effectiveness of integrating non-structural measures like early warning systems with structural measures like levees in reducing loss of life during inland flooding (medium confidence). In addition, comprehensive disaster risk management strategies, climate services, and social safety nets have proven broadly applicable across sectors, enhancing overall resilience. Looking towards mitigation, the report identifies numerous strategies –notably solar and wind energy, electrification of urban systems, urban green infrastructure, energy efficiency measures, demand-side management, improved forest and crop/grassland management, and reduced food waste and loss – as not only technically feasible and increasingly cost-effective but also widely supported by the public (IPCC, 2023).

2.2 *IPCC AR7 Special Report* Climate Change and Cities

The IPCC, having initiated its seventh cycle in 2023, has identified the need for a special report on the topic of cities in climate change, specifically. This decision is grounded in several key factors. To begin with, the topic is highly relevant to both WG2 and WG3. WG2 focuses on assessing the unique impacts of climate change on urban areas, including the combined effects of climate change and urban heat islands, as well as urban vulnerabilities, resilience, and adaptation strategies. WG3, on the other hand, examines mitigation opportunities at the city level and their integration with broader urban processes and priorities (IPCC, n.d.).

Furthermore, while the AR5 Report (IPCC, 2014) was the first to acknowledge the urban scale as a significant unit of analysis, it left considerable gaps. The sectoral approach employed by WG3 in AR5 proved inadequate for assessing the complex and interconnected nature of urban processes and opportunities. The AR5 also failed to fully capture the wide range of integrative approaches at the urban scale that offer more than the sum of individual sectoral options. A dedicated special report on cities can provide a more tailored analytical framework, effectively capturing and portraying the full range of urban adaptation and mitigation opportunities. In addition, the existing literature and practical action on urban climate issues have expanded rapidly, yet single chapters in WG reports can only scratch the surface of these complex topics. Integrating adaptation and mitigation strategies at the urban level is crucial to avoid counterproductive actions and maximise synergistic opportunities, a goal that cannot be achieved by addressing the urban scale in separate volumes. Critically, while recent assessments have touched upon cities and climate change, none have comprehensively addressed the intersection of these issues in a manner that meets the needs of the IPCC's key target audiences.

Of particular significance is the increasing recognition of cities as potential hubs for flexible and rapid climate action, which has led to a surge in both practical initiatives and research on this topic, further underscoring the need for a focused assessment. The proposed Special Report will not only complement the AR6 assessment but also offer a unique lens through which to analyse urban climate change impacts and opportunities for action. This will provide the IPCC audience with diverse insights and potentially catalyse different types of climate action, reaching a wider range of stakeholders. Moreover, the detailed assessment of the urban scale in the Special Report will alleviate the pressure on AR6 to be both comprehensive and balanced in its treatment of urban issues, allowing for a more in-depth analysis of key topics. Given the rapidly expanding knowledge base, the limited prior attention given to this critical scale, and the growing emphasis on city-level climate initiatives, this Special Report is both timely and essential. It is poised to serve as a vital bridge between WG2 and WG2, fostering integration and facilitating more effective, holistic approaches to addressing climate change in the urban context.

In 2016, the IPCC decided to enhance the integration of urban issues within the AR6 assessment framework. This includes a stronger focus on the impacts of climate change on cities, their unique adaptation and mitigation opportunities, and a more robust consideration of cities within the broader context of regional issues and human settlements. The IPCC further emphasised the importance of engaging urban practitioners in this process. In addition, the decision affirmed the inclusion of a Special Report on climate change and cities in the AR7 cycle, recognising the growing significance of urban areas in the global climate change discourse. IPCC also expressed interest in collaborating with academia, urban practitioners, and relevant scientific bodies and agencies to organise an international scientific conference on climate change and cities early in the AR6 cycle. The aim of this conference would be to stimulate scientific reports and peer-reviewed publications on this critical topic, thus enriching the knowledge base and informing future assessments and policy decisions.

2.3 IPCC Global Research and Action Agenda on Cities and Climate Change Science

The "Global Research and Action Agenda on Cities and Climate Change Science" (IPCC, 2024) highlights the potential for cities to be significant catalysts in implementing international agreements such as the Paris Agreement, the 2030 Sustainable Development Agenda, the New Urban Agenda, and the Sendai Framework for Disaster Risk Reduction. City-level adaptation and mitigation actions will be crucial in supporting national efforts to meet these commitments, particularly given the projected increase in urban populations, with 68% of the world's population expected to reside in cities by 2050 (UN DESA, 2018). The Agenda aims to assist national governments, local and municipal authorities, researchers, scientists, planning and design communities, private sector enterprises, international organisations, and civil society, including Indigenous people, in developing evidence-based research and knowledge to support effective climate action in cities. It outlines key research areas that will inform policy development for urban climate action (IPCC, 2024).

Recognising the critical role of cities in the global response to climate change, the IPCC approved a proposal for the co-sponsored International Conference on Climate Change and Cities (renamed and branded as the Cities and Climate Change Science Conference) at its 44th Session in Bangkok. This landmark conference, held in Edmonton, Canada, from 5 to 7 March 2018, brought together over 700 participants to assess existing knowledge and identify critical research gaps in the field of cities and climate change. Insights from this conference, which encompassed cities of diverse sizes, growth patterns, geographies, and contexts, informed the development of this Global Research and Action Agenda on Cities and Climate Change Science. The document is structured into three sections: an exploration of cross-cutting issues and knowledge gaps, an examination of key topical research areas, and a presentation of suggested approaches for implementing the Research and Action Agenda (IPCC, 2024). The structure of the Research and Action Agenda is illustrated in Figure 2.1.

The first section of the Agenda delves into cross-cutting issues and knowledge gaps that span various urban systems and scales, emphasising the need for collaboration and co-production of knowledge to tackle complex challenges (IPCC, 2024). The second section focuses on key topical research areas, ranging from the built and natural environment to social, economic, and governance dimensions, reflecting the multifaceted nature of urban climate action. Lastly, the Agenda presents suggested approaches for implementing effective climate solutions in cities, with a particular emphasis on systems thinking and financial mechanisms. While intended to be broadly applicable,

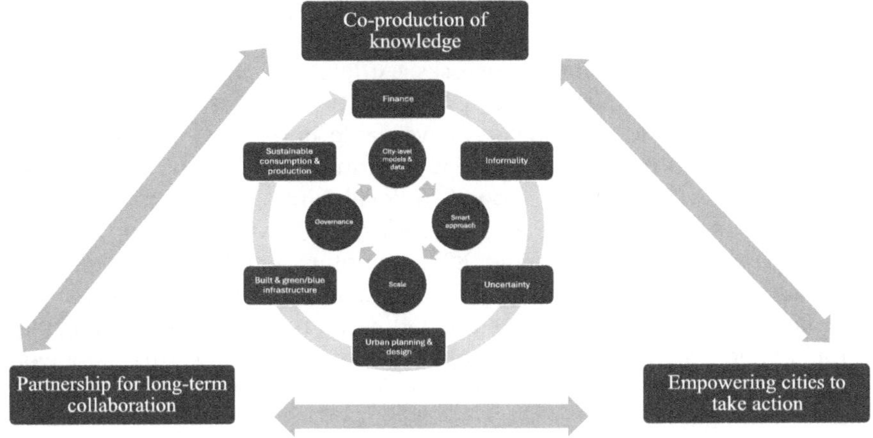

Figure 2.1 Pathways for climate adaptation and mitigation in cities.
(*Source:* IPCC, 2024)

the agenda acknowledges that certain aspects may be more relevant to specific urban contexts, reflecting the diversity of cities worldwide.

2.4 UNFCCC Decision Under Paris Agreement

From 30 November to 12 December 2023, the COP28 conference in Dubai marked the completion of the first GST under the Paris Agreement (UNFCCC, 2015). The Conference of the Parties serving as the meeting of the Parties to the Paris Agreement (CMA) decision on the first GST was adopted by all parties. This landmark agreement, adopted at COP21 in 2015 (Paris) and entering into force in November 2016, replaced the Kyoto Protocol and established a new international framework for GHG emissions reduction beyond 2020, encompassing both developed and developing countries. The Paris Agreement outlines several key provisions aimed at addressing climate change. Foremost, it establishes a long-term goal to limit global warming to well below 2°C above pre-industrial levels, while pursuing efforts to limit the increase to 1.5°C. This includes achieving a balance between anthropogenic GHG emissions and removals in the second half of the century. The Agreement requires all parties, including major emitting nations, to submit and regularly update nationally determined contributions (NDCs) outlining their GHG emissions reduction targets. These NDCs are to be updated every five years, promoting continuous improvement in climate action. In addition, it promotes the use of market mechanisms, such as the Joint Crediting Mechanism (JCM), to incentivise emission reductions. These mechanisms aim to create economic incentives for countries and businesses to reduce emissions while fostering sustainable development.

Moreover, the Paris Agreement recognises the importance of adaptation by mandating that parties set long-term adaptation goals (Article 7.1), implement national adaptation planning processes (Article 7.9), and submit periodic adaptation communications (Article 7.10). This emphasises the need for proactive measures to address the unavoidable impacts of climate change. To support these efforts, the Agreement reaffirms the obligation of developed countries to provide financial support to developing countries for mitigation and adaptation efforts (Article 9). It also encourages voluntary contributions from developing countries (Article 9.2), promoting a collaborative approach to climate finance. In addition, the Agreement establishes a transparency framework (Article 13) requiring all parties to report on their mitigation and adaptation actions, subject to technical expert review. This framework ensures accountability and transparency in the implementation of national climate commitments.

Finally, the Paris Agreement establishes a GST mechanism to assess collective progress towards achieving the Agreement's objectives every five years. The GST serves as a critical review process, allowing parties to evaluate their collective efforts and identify areas for improvement. Within the framework of the Paris Agreement, the GST serves to periodically assess collective progress towards achieving the Agreement's objectives and long-term goals, with each assessment occurring every five years. Regarding the scope of the GST, Decision 19 CMA1 (UNFCCC, 2019) established a technical dialogue to take stock of the implementation of the Paris Agreement. This process assesses collective progress towards the Agreement's purpose and long-term goals, as outlined in Article 2, paragraph 1(a-c) of the Agreement, focusing on the thematic areas of mitigation, adaptation, and means of implementation and support (UNFCCC, 2015). In addition, the GST may consider efforts related to addressing the social and economic consequences of response measures and averting, minimising, and addressing loss and damage associated with the adverse effects of climate change.

The outputs of this component of the GST should identify opportunities and challenges in enhancing action and support, outline possible measures and good practices, and summarise key political messages, including recommendations for strengthening action and support, including recommendations arising from COP28. The outcome of the first GST was adopted as one of the Decisions of the CMA5 session, held in the United Arab Emirates from 30 November to 13 December 2023, which emphasises the important role of cities (UNFCCC, 2024).

Paragraph 158 of the Decision acknowledges the significant role and active engagement of non-Party stakeholders, notably cities and subnational authorities, in supporting Parties to the Paris Agreement (UNFCCC, 2024). These stakeholders have contributed to substantial collective progress towards achieving the Agreement's temperature goal, addressing and responding to climate change, and enhancing ambition, including through participation in other relevant intergovernmental processes. Paragraph 162 further encourages international cooperation and the exchange of views and experience among non-Party stakeholders at all levels, including local and subnational, through activities such as joint research, personnel training, practical projects, technical exchanges, project investment, and standards cooperation (UNFCCC, 2024).

2.5 G7 Roundtable on Subnational Climate Actions

Cities account for approximately 70% of global GHG emissions, placing subnational governments at the forefront of global efforts to mitigate and adapt to climate change. Collaborative action across subnational, national, and international levels, in conjunction with civil society and the private sector, is crucial for enabling and accelerating the transition towards sustainable and climate-resilient development pathways. To this end, the G7 Sapporo Ministers' Meeting on Climate, Energy and Environment, held on 15–16 April 2023, resolved to promote climate action by fostering closer cooperation with subnational governments. This includes facilitating international city-to-city collaboration and engaging with G7 ministers responsible for urban development on net-zero and resilience agendas. As a result, the "G7 Roundtable on Subnational Climate Actions" was established to facilitate the sharing of national policies and programmes that promote subnational climate action, both domestically and internationally. The Roundtable aims to identify co-benefits and explore opportunities for international cooperation in this critical area.

The G7 Roundtable on Subnational Climate Actions was officially convened by the MOEJ on 5–6 October 2023. The event was co-hosted by the Ministry of Land, Infrastructure, Transport and Tourism of Japan, the Institute for Global Environmental Strategies, and Urban 7 represented by the ICLEI – Local Governments for Sustainability. The participants included policymakers from ministries of both the environment and urban development of G7 members (Japan [G7 Presidency], Canada, European Union [EU], France, Germany, Italy, the United Kingdom, and the United States). The Roundtable group facilitated in-depth discussions among policymakers from G7 member states regarding policies and programmes that encourage subnational climate action, both domestically and internationally. It underscored the pivotal role of cities in achieving deep emission reductions, emphasising the importance of subnational governments in accelerating the transition towards net-zero development pathways. It highlighted the need for close collaboration between national and subnational governments, including through the support of international city networks, to facilitate knowledge sharing and effective policy implementation. Moving forward, the event explored potential collaborations with G7 ministers responsible for urban development.

At the session held in October 2023, successful case studies were shared regarding national support policies and programmes for promoting subnational climate actions, both within and beyond G7 countries. Through candid breakout sessions with G7 representatives, two key areas were addressed:

first, the enhancement of national support policies and programmes, including the leveraging of synergies, mainstreaming of climate actions into urban development and management, and the utilisation of digital technology. The second area is the advancement of international cooperation by G7 national governments to support subnational climate actions in non-G7 countries, exploring opportunities for collaborative initiatives among G7 nations and the development of integrated smart cities with user benefits. The outcomes of the Roundtable, consolidated into a Summary Report, are intended to be shared by the MOEJ with non-G7 countries in various international forums. Notably, the Report was presented at the COP28 conference to the UNFCCC.

2.6 The CDP-ICLEI Unified Reporting System

Every year, cities voluntarily report their climate and environmental performance data to the Carbon Climate Registry, also known as the CDP-ICLEI Unified Reporting System. This system, managed by ICLEI – Local Governments for Sustainability, encourages participants to gain a deeper understanding of the climate risks associated with urban-related GHG emissions by monitoring them annually.

According to the "Carbon" Climate Registry Report entitled *Multilevel Climate Action: The Path to 1.5 Degrees* (ICLEI, 2018), 861 cities globally disclosed their data through the CDP-ICLEI Unified Reporting System. Of these, 521 cities (61%), representing 73 countries and approximately 8% of the global population, reported implementing climate change mitigation actions from a list of 48 options provided by the system. These cities also identified anticipated co-benefits resulting from these actions, based on a list provided within the reporting framework.

The objective of this Report was to assess how cities identify and leverage co-benefits to support their climate action initiatives (ICLEI, 2018). It analyses the climate mitigation actions and associated co-benefits reported by cities from 2018 to 2019 and includes case studies of cities actively integrating co-benefits into their climate action planning. In addition, the Report serves as a resource guide, detailing tools and information available to cities seeking to embed co-benefits into their climate action strategies (ICLEI, 2018). Relevant to the focus of this chapter, Figure 2.2 illustrates the most frequently reported mitigation strategies implemented by the participating cities.

The most frequently reported mitigation actions undertaken by cities participating in the CDP-ICLEI Unified Reporting System, as shown in the figure, predominantly focus on improving energy efficiency in buildings and transitioning to renewable energy sources. This aligns with the IPCC's AR6 Synthesis Report (2023), which underscores the importance of transitioning towards renewable energy to mitigate climate change. However, the city-level data reveals a notable absence of emphasis on ecosystem-based adaptation approaches, such as urban greening and wetland restoration, which the AR6 report identifies as effective strategies for mitigating specific climate risks like

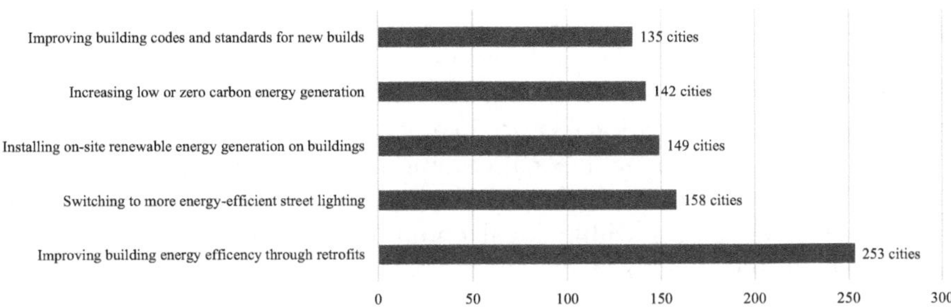

Figure 2.2 The co-benefits of climate action accelerating city-level ambition
(*Source:* CDP, 2020)

floods and urban heat island effects. This discrepancy suggests a potential gap in urban climate action planning, with cities prioritising energy-focused mitigation over ecosystem-based adaptation measures, despite the latter's recognised effectiveness in addressing certain climate vulnerabilities.

3. Local Decarbonisation and International Cooperation in Japan

This section focuses on the ongoing situation in Japan in the context of local decarbonisation and international cooperation. Subsections 3.1 and 3.2 introduce policies and actions in Japan related to local decarbonisation, while subsection 3.3 describes successful case studies and good practices led by Japan in the context of international cooperation.

3.1 Mitigation: Policies and Initiatives in Japan from the Context of Local Decarbonisation

To achieve carbon neutrality in 2050, the Government of Japan has decided to reduce GHG emissions by 46% by 2030. Subsequently, local government leaders and ministers discussed decarbonisation measures in regional initiatives and the "Regional Decarbonisation Roadmap" was formulated in June 2021. The Roadmap prioritises problem-solving and decarbonising regional development, starting with 100 model locations nationwide that aim to achieve the decarbonisation goals by 2030. So far, 46 decarbonisation leading areas have been selected. The MOEJ will call for applications semi-annually until 2025 (MOEJ, 2021).

According to the Roadmap's contents, local entities are encouraged to take the lead in implementing measures such as self-consumption, solar power generation, and energy-saving initiatives in homes and buildings. The MOEJ created advanced models of decarbonisation in diverse regions such as urban, agricultural, and tourist areas. The goal is to expand these efforts to national and overseas markets, triggering the "decarbonisation domino effect" that fosters sustainable and affluent regions throughout the country

(MOEJ, 2021). Initiatives outlined in the Roadmap include utilising human resources, information, and finance, as well as innovating lifestyles and rules to promote decarbonisation. The ultimate vision is to create sustainable and affluent regions by solving local issues in a decarbonising manner, expanding renewable energy that benefits local communities, and encouraging electric vehicle use to enrich local economies. Additionally, it also envisions a future where money circulates within local communities, and residents enjoy a sustainable lifestyle, thereby achieving the national goal of net-zero emissions by 2050 (MOEJ, 2021).

As support from the national government, the MOEJ provides three types of support to local governments (MOEJ, 2021): capacity development for both local governments and the private sector, knowledge support including renewable energy potential and local economic analysis, and financial support to empower local governments and attract private sector investment for decarbonisation projects.

3.2 Adaptation: Policies and Initiatives in Japan in the Context of Local Decarbonisation

This subsection presents the Japanese 2018 Climate Change Adaptation Act and materials on adaptation submitted by Japan to the UNFCCC in October 2021.

3.2.1 Climate Change Adaptation Act

The Climate Change Adaptation Act was established in Japan in 2018. The Act outlines the roles of national and local governments, the private sector, and citizens in promoting climate change adaptation. It mandates the national government to conduct climate impact assessments every five years, ensuring that national measures are based on scientific evidence. Furthermore, local governments are required to formulate local adaptation plans and designate local adaptation centres to address region-specific climate vulnerabilities (see Figure 2.3).

The Climate Change Adaptation Act establishes a four-pillar framework for climate change adaptation: (1) a comprehensive plan defining roles for various stakeholders and mandating national climate impact assessments every five years; (2) an information platform (A-PLAT) operated by the National Institute for Environmental Studies (NIES) to provide technical advice and data to local governments; (3) adaptation measures at the local level, including the formulation of local plans, the designation of adaptation centres, and the organisation of regional councils, and (4) international actions promoting cooperation and fostering adaptation businesses (Cabinet of the Government of Japan, 2018). The overarching goal is to foster effective adaptation measures across various sectors, ranging from agriculture and natural ecosystems to human health and urban life, based on reliable scientific information and future projections.

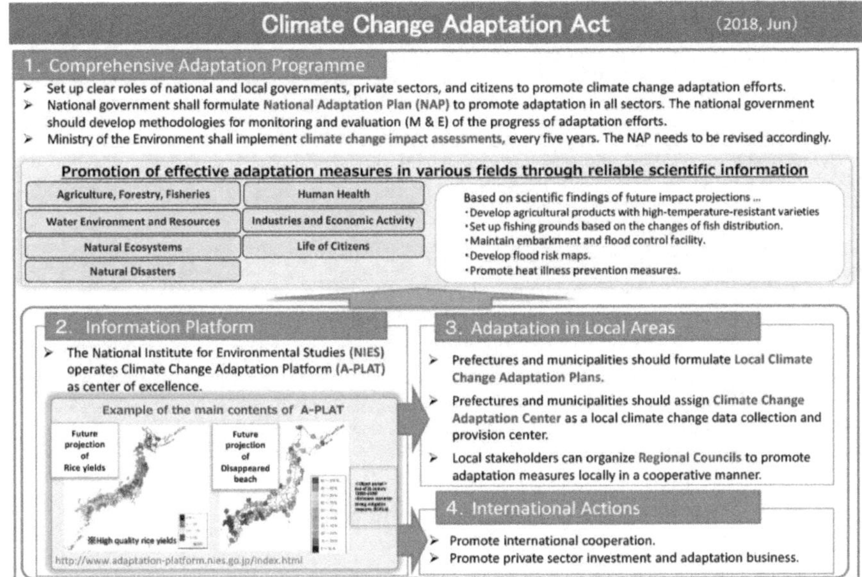

Figure 2.3 The conceptual image of Climate Change Adaptation Act
(*Source:* MOEJ, 2023)

3.2.2 Japan's Adaptation Communication to the UNFCCC

In compliance with Article 7 of the Paris Agreement (United Nations, 2015), which mandates each party to submit and periodically update an adaptation communication detailing their climate change priorities, implementation strategies, support needs, plans, and actions, Japan submitted its Adaptation Communication to the UNFCCC in October 2021, establishing 66 key performance indicators (KPIs) to track climate change adaptation and mitigation progress. Notably, the Japanese government opted to focus solely on adaptation in this communication, rather than incorporating it into the NDC. As illustrated in Figure 2.4, Japan's Adaptation Communication highlights seven "noteworthy efforts", with effort number 5 as, "Efforts for Adaptation at the Local Level," indicating that 46 out of 47 prefectures have developed local climate change adaptation plans.

The Adaptation Communication also highlights the establishment of the Climate Change Adaptation Promotion Council, a body tasked with coordinating adaptation efforts at the national level. In terms of international cooperation, Japan is actively contributing through the Asia-Pacific Climate Change Adaptation Information Platform (AP-PLAT) and the provision of climate finance to support adaptation in other countries. Furthermore, the communication underscores Japan's commitment to individual-level adaptation measures, such as addressing heatstroke and weather-related disasters.

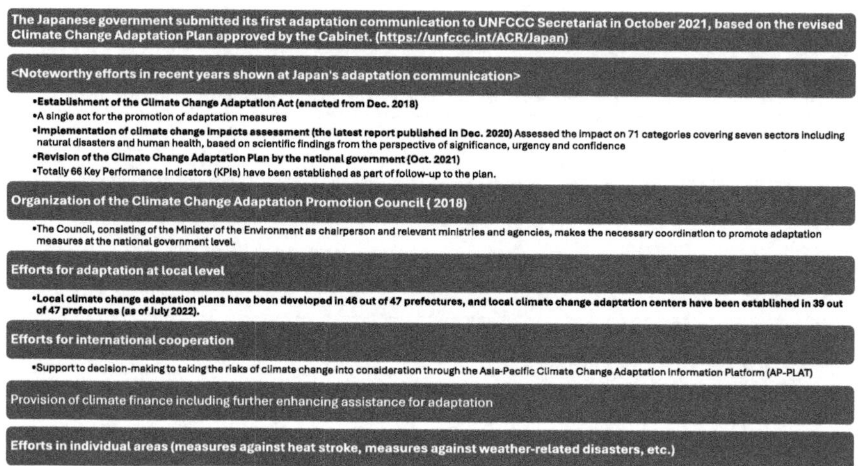

Figure 2.4 Outline of Japan's Adaptation Communication
(*Source:* UNFCCC, 2023)

3.3 *Successful Case Studies and Good Practices Led by Japan in the Context of International Cooperation*

This subsection provides success cases and good practices led by Japan from the context of international cooperation through City-to-City Collaboration Programme (C3P) and Clean City Partnership Programme (C2P2).

3.3.1 City-to-City Collaboration Programme

The MOEJ is implementing a programme to facilitate city-to-city collaboration between Japanese and international cities. This initiative aims to promote knowledge sharing and experience exchange for decarbonisation, in partnership with private solution providers. The objectives and guidelines for this programme are detailed in the document "Creating Sustainable, Zero-Carbon Societies Through City-to-City Collaboration" (MOEJ, 2021). The Japanese Ministry recognises that due to the concentration of infrastructure in cities, "the introduction and development of superior zero-carbon technologies, products, and systems in these facilities will not only contribute to the decarbonization of cities, but also generate various co-benefits, such as improving the environment and energy supply in cities" (MOEJ, n.d.).

Within the C3P, as shown in Figure 2.5, Japanese private companies, under contract with the MOEJ and local municipalities, collaborate on developing zero-/low-carbon projects in partner cities within developing countries. This collaboration involves establishing foundational systems to facilitate zero-carbon societies and enhancing the capacity of municipal staff in these partner cities. The programme is mutually beneficial, fostering zero-/

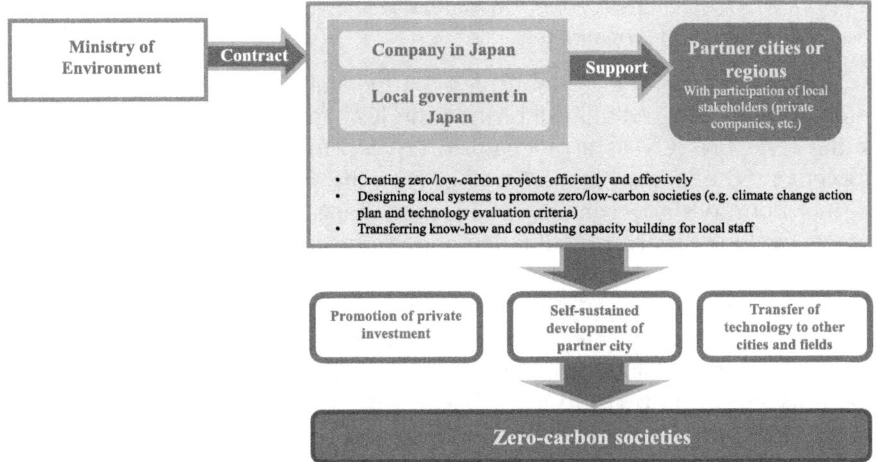

Figure 2.5 Conceptual diagram of the City-to-City Collaboration Programme

(*Source:* MOEJ, 2023)

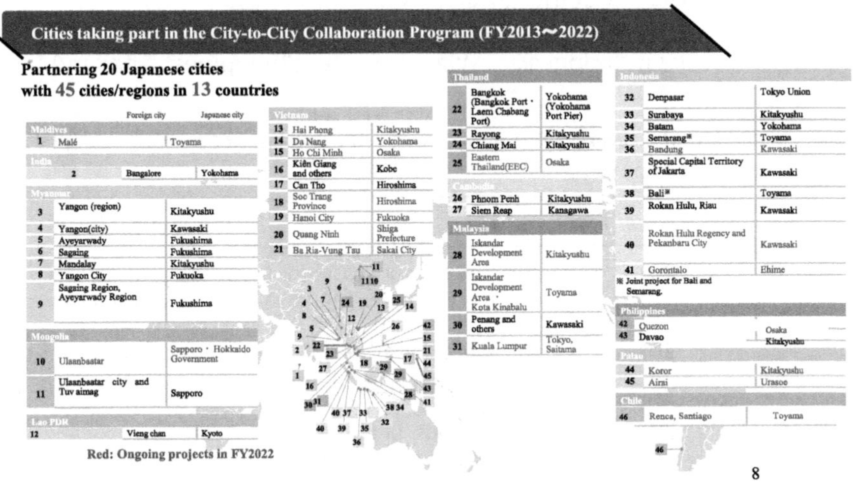

Figure 2.6 Cities participating in the City-to-City Collaboration Programme (FY2013 to 2023)

(*Source:* MOEJ, 2023)

low-carbon and self-sustaining development in partner cities, while also stimulating private investment within Japan (MOEJ, 2021). The MOEJ facilitates partnerships between 20 Japanese subnational governments and 49 subnational governments across 13 countries, as illustrated in Figure 2.6.

3.3.2 JCM Model Projects Formulated in the Framework of City-to-City Collaboration Programme

As the next step after establishing partnerships between Japanese and international cities, MOEJ established the JCM formulated in the framework of the C3P. The JCM is a mechanism created in 2014, jointly created and implemented under an agreement between the Government of Japan and a partner country's government to achieve Japan's GHG emissions reduction targets by "quantitatively evaluating and understanding Japan's contributions to the reduction/removal of GHG emissions achieved through the spread of superior zero-/low-carbon technologies (technologies/products, systems, services, infrastructure, etc.) and the implementation of policies that can lead to a reduction in GHG emissions in developing countries" (MOEJ, 2021). The JCM also contributes to the achievement of the ultimate objective of the UNFCCC by promoting actions to reduce and remove GHG emissions on a global scale. Japan, as of February 2020 (MOEJ, 2021), was implementing the JCM with 17 countries: Mongolia, Bangladesh, Ethiopia, Kenya, Maldives, Viet Nam, Laos, Indonesia, Costa Rica, Palau, Cambodia, Mexico, Saudi Arabia, Chile, Myanmar, Thailand, and the Philippines. Notably, with regard to the regional collaboration, this model project has disseminated 24 projects in 6 Southeast Asian countries (MOEJ, 2021).

As illustrated in Table 2.1, the participating Southeast Asian countries include Cambodia, Indonesia, Myanmar, the Philippines, Thailand, and

Table 2.1 JCM model projects formulated in the framework of City-to-City Collaboration Programme

Country	*Fiscal Year*	*Project*	*Partners*
Myanmar	2015	Introduction of Waste to Energy Plant in Yangon City	Yangon City – Kawasaki City
	2016	Introduction of Energy Saving Brewing Systems to Beer Factory	Yangon City – Kawasaki City
	2016	Introduction of High-Efficiency Boiler in Instant Noodle Factory	Yangon City – Kawasaki City
	2016	Rice Husk Power Generation in Rice Mill Factory in Ayeyarwady	Ayeyarwady Region – Fukushima City
Thailand	2016	Introduction of 12 MW Power Generation System Waste Heat Recovery for Cement Plant	Rayong Province – Kitakyushu City
	2021	Introduction of High-Efficiency Boiler to Garment Factory	Eastern Thailand – Osaka City

(*Continued*)

Table 2.1 (Continued)

Country	*Fiscal Year*	*Project*	*Partners*
The Philippines	2021	Introduction of Energy Saving Air Conditioning System to Quezon City Hall Compound	Quezon – Osaka City
Vietnam	2014	Eco-Driving by Utilising Digital Tachograph System	Ho Chi Minh City – Osaka City
	2015	Introduction of Solar Photovoltaic System at Shopping Mall in Ho Chi Minh	Ho Chi Minh City – Osaka City
	2015	Energy Saving in Factories with Air-Conditioning Control System	Ho Chi Minh City – Osaka City
	2016	Introduction of High-Efficiency Water Pumps in Da Nang City	Da Nang City – Yokohama City
	2019	Introduction of High-Efficiency Air Conditioning System and Air-Cooled Chillers to Office Buildings	Ho Chi Minh City – Osaka City
	2020	Introduction of High-Efficiency Boiler System to Food Factor	Ho Chi Minh City – Osaka City
	2020	Introduction of High-Efficiency Air-Conditioning System to Hotel in Ho Chi Minh City	Ho Chi Minh City – Osaka City
	2021	Introduction of High Efficiency LED Lighting to Office Building in Ho Chi Minh City	Ho Chi Minh City – Osaka City
	2021	Introduction of 9.8 MW Rooftop Solar Power System in Industrial Park	Ho Chi Minh City – Osaka City
	2022	Mini Hydropower Plant Project	Hanoi City – Fukuoka Prefecture
	2022	Introduction of 0.4 MW Rooftop Solar Power System to Aluminum Wheel Manufacturing Factory	Ho Chi Minh City – Osaka City
Cambodia	2016	Introduction of 1 MW Solar Power System and High-Efficiency Centrifugal Chiller in Large Shopping Mall	Phnom Penh City – Kitakyushu City

(*Continued*)

Table 2.1 (Continued)

Country	*Fiscal Year*	*Project*	*Partners*
	2015	Energy Saving for Air-Conditioning at Shopping Mall with High-Efficiency Centrifugal Chiller	Surabaya City – Kitakyushu City
	2015	Energy Saving for Industrial Park with Smart LED Street Lighting System	Surabaya City – Kitakyushu City
Indonesia	2018	Introduction of CNG-Diesel Hybrid Equipment to Public Bus in Semarang	Semarang City – Toyama City

(*Source:* MOEJ, 2023)

Vietnam. The JCM projects primarily focused on energy efficiency and renewable energy integration, encompassing initiatives such as the introduction of waste-to-energy plants, high-efficiency boilers and chillers, solar PV systems in various settings (shopping malls, industrial parks, rooftops), and energy-saving technologies in factories and public buildings. Notably, a project in Semarang, Indonesia, introduced CNG[2]-diesel hybrid equipment to public buses, demonstrating a commitment to sustainable transportation solutions. The diverse range of projects implemented across these countries reflects a concerted effort to reduce GHG emissions and promote sustainable development in the region. The following subsection will therefore discuss one of the successful case studies of collaboration within the CP3 framework.

3.4 Collaboration Between Yokohama City and Bangkok Metropolitan Administration

The Japan International Cooperation Agency (JICA) is a government-affiliated financial institution that provides funding and financing for projects established within the JCM under the C3P framework. JICA's cooperation with the Bangkok Metropolitan Administration (BMA) led to the development of the Master Plan on Climate Change, which focuses on five key areas: promoting environmentally sustainable transport, improving energy efficiency, enhancing waste management, implementing green urban planning, and developing adaptation plans. The BMA has successfully reduced GHG emissions and improved air quality through initiatives such as public transport development, energy efficiency improvements, and waste reduction (MOEJ, 2023). Yokohama City actively participated in the C3P collaboration with Bangkok by providing expertise and training to the BMA. Furthermore, Yokohama City and the BMA have continued to collaborate on the implementation of the Master Plan with continuous financial support from the MOEJ (MOEJ, 2023).

A notable initiative within the C3P between Japan and Thailand is the Smart Ports project (see Figure 2.7). The Yokohama Port Corporation, in collaboration

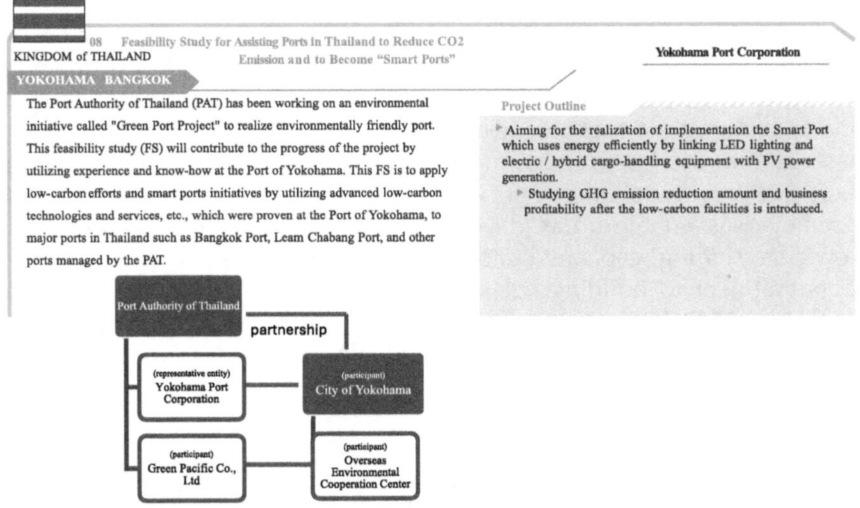

Figure 2.7 Conceptual image of the success of city-to-city collaboration between Yokohama City and the BMA

(*Source:* MOEJ, 2023)

with the Port Authority of Thailand (PAT), the City of Yokohama, Green Pacific Co. Ltd., and the Overseas Environmental Cooperation Centre, conducted a feasibility study to assist major Thai ports, including Bangkok and Leam Chabang, in reducing CO_2 emissions and transitioning towards more sustainable operations. The study aimed to leverage the expertise and experience gained at the Port of Yokohama in implementing low-carbon technologies and smart port initiatives. The project outline focuses on improving energy efficiency by integrating LED lighting and electric/hybrid cargo-handling equipment with photovoltaic (PV) power generation. The study also assessed the potential reduction in GHG emissions and the economic benefits associated with implementing these low-carbon technologies. This collaborative effort exemplifies a significant step towards more sustainable and environmentally friendly port operations in Thailand.

3.4.1 Clean City Partnership Programme

The C2P2, launched in February 2023 by the MOEJ and the JICA, aims to address multifaceted challenges faced by cities worldwide. By collaborating with Japanese local governments, private companies, financial institutions, and multilateral development banks, C2P2 aims to provide comprehensive and synergistic support to partner cities overseas. This support targets urban challenges including climate change mitigation and adaptation, environmental pollution reduction, the promotion of a circular economy, and the reversal of biodiversity loss (JICA, 2023). To promote C2P2, a seminar titled *Urban Agenda on Climate Change, Pollution and Biodiversity Loss*, focusing on the C2P2, was held at COP28 conference. The event brought together academics, specialists, and stakeholders to discuss the programme's focus areas, which include climate action, circular economy, pollution control, and biodiversity conservation. By fostering partnerships between Japanese and overseas cities, C2P2 aims to implement innovative technologies and approaches to achieve sustainable urban development. The programme's emphasis on knowledge sharing, capacity building, and collaborative action underscores its commitment to building resilient and environmentally conscious cities worldwide. As C2P2 continues to evolve, it is poised to play a crucial role in driving sustainable urban development globally.

4. Discussion

The assessment of various initiatives and reports, including the IPCC AR6, the Paris Agreement, the G7 Roundtable on Subnational Climate Actions, and the CDP-ICLEI Unified Reporting System, reveals a multifaceted and evolving landscape of urban climate action. These initiatives converge on the recognition of cities as crucial actors in climate change mitigation and adaptation, underscoring the necessity for integrated and comprehensive approaches that address both current challenges and future uncertainties.

The IPCC reports highlight the significant potential of cities to contribute to global emissions reduction through strategies such as energy efficiency, electrification, and nature-based solutions. They also emphasise the importance of addressing consumption-based emissions and collaborating with diverse stakeholders to achieve net-zero goals. The Paris Agreement, with its emphasis on NDCs and GSTs, provides a framework for integrating urban climate action into broader national and international efforts.

The G7 Roundtable on Subnational Climate Actions and the CDP-ICLEI Unified Reporting System demonstrate the growing momentum of city-level climate initiatives, showcasing innovative approaches and collaborative efforts to accelerate the transition towards sustainable and resilient urban environments. The emphasis on knowledge sharing, capacity building, and international cooperation highlights the recognition that climate action in cities requires a multi-level governance approach and the involvement of diverse stakeholders. However, despite this progress, the analysis also reveals a gap between the scientific understanding of effective climate strategies and their actual implementation in cities. While cities are prioritising energy-focused mitigation, there is a need for greater emphasis on ecosystem-based adaptation approaches, which are crucial for addressing specific climate vulnerabilities. Furthermore, the integration of climate action into broader urban planning and development processes remains a challenge, requiring a more holistic and systemic approach. Looking towards the future, the upcoming IPCC Special Report on Climate Change and Cities promises to further deepen our understanding of urban climate dynamics and provide a more comprehensive assessment of the challenges and opportunities. The continued engagement of cities in global climate initiatives, such as the COP conferences and the G7 process, will be crucial for translating scientific knowledge into actionable policies and ensuring that cities play a leading role in the global transition towards a sustainable and resilient future.

In terms of adaptation, domestically, the Regional Decarbonization Roadmap and the Climate Change Adaptation Act provide comprehensive frameworks for achieving net-zero emissions by 2050 and enhancing resilience to climate impacts. These initiatives emphasise collaboration between national and local governments, knowledge sharing, capacity building, and financial incentives to drive decarbonisation efforts across various sectors. Internationally, Japan's commitment is evident through its Adaptation Communication to the UNFCCC, highlighting significant progress in local adaptation planning and showcasing noteworthy efforts in climate resilience. In addition, the C3P and the C2P2 exemplify Japan's dedication to fostering international cooperation. These programmes facilitate knowledge exchange, technology transfer, and capacity building between Japanese cities and their counterparts in developing countries, focusing on areas such as renewable energy, energy efficiency, waste management, and smart city development. The success of initiatives like the JCM and the collaborative efforts between cities like Yokohama and Bangkok demonstrate the tangible impact of these programmes. By leveraging its expertise and resources, Japan is playing a crucial role in driving sustainable urban development and climate action on a global scale.

However, challenges remain, including the need for continued investment in research and development, the scaling up of successful pilot projects, and ensuring equitable distribution of benefits from these initiatives. The evolving nature of climate change necessitates continuous monitoring, evaluation, and adaptation of strategies to ensure their long-term effectiveness. Nevertheless, Japan's comprehensive and multifaceted approach to climate change mitigation and adaptation, both domestically and internationally, serves as a model for other countries. The emphasis on collaboration, knowledge sharing, and capacity building is crucial for achieving global climate goals. While challenges persist, the progress made thus far demonstrates the potential of these initiatives to create a more sustainable and resilient future for cities worldwide.

5. Conclusion

This chapter explored the multifaceted landscape of urban climate action, highlighting the growing recognition of cities as critical players in both mitigation and adaptation. International initiatives, such as the Paris Agreement and the G7 Roundtable on Subnational Climate Actions, are actively promoting city-level engagement through knowledge sharing, capacity building, and financial mechanisms. These efforts underscore a global commitment to harnessing the potential of cities in combating climate change.

Japan's comprehensive approach to climate action, encompassing both local decarbonisation and international cooperation, serves as a compelling model. The country's domestic policies, such as the Regional Decarbonisation Roadmap and the Climate Change Adaptation Act, have established a robust framework for achieving net-zero emissions and enhancing resilience, supported by the provisions of the Green Transformation (GX) Promotion Act (Cabinet of the Government of Japan, 2023). Furthermore, Japan's international initiatives, notably the C3P and the C2P2, exemplify a commitment to sharing expertise and resources with developing countries, fostering a global network of sustainable cities.

However, the challenge remains to bridge the gap between scientific understanding and practical implementation, particularly in scaling up ecosystem-based adaptation measures alongside energy-focused mitigation. While the momentum for urban climate action is growing, continued investment in research and development, equitable resource distribution, and adaptive strategies are essential for long-term success. As cities continue to account for a substantial share of global GHG emissions, estimated at 67–72%, the urgency for comprehensive urban climate action intensifies. Smart urban planning, waste management, and the transition towards low-carbon energy sources are crucial steps in reducing emissions. Furthermore, investing in energy-efficient infrastructure, promoting non-motorised transport, and integrating nature-based solutions will be essential in building climate-resilient cities.

The future of global and local climate initiatives hinges on sustained collaboration between national governments, local authorities, and international organisations, along with the active participation of civil society and the private sector. It is through such collaborative efforts that cities can fully realise their potential as drivers of innovation and transformative action in the fight against climate change.

Notes

1 Global net anthropogenic GHG emissions equivalent.
2 Compressed natural gas.

References

Asia-Pacific Climate Change Adaptation Information Platform (AP-PLAT) (n.d.) "Get Started", National Institute for Environmental Studies. Available at: https://ap-plat.nies.go.jp/get_started/index.html (Accessed: 28 June 2024).

Cabinet of the Government of Japan (2018) "Climate Change Adaptation Act". Available at: https://www.env.go.jp/content/900451276.pdf (Accessed: 29 June 2024).

Cabinet of the Government of Japan (2023) "Overview of Japan's Green Transformation (GX) Promotion Act". Available at: https://grjapan.com/sites/default/files/content/articles/files/gr_japan_overview_of_gx_plans_january_2023.pdf (Accessed: 29 June 2024).

Carbon Disclosure Project (CDP) (2020) "The Co-Benefits of Climate Action: Accelerating City-Level Ambition". Available at: https://cdn.cdp.net/cdp-production/

cms/reports/documents/000/005/329/original/CDP_Co-benefits_analysis.pdf?1597235231 (Accessed: 28 June 2024).
Government of Japan (2021a) "Adaptation Communication Pursuant to Article 7, Paragraph 10 of the Paris Agreement". Available at: https://unfccc.int/sites/default/files/resource/JAPAN_adaptation_communication.pdf (Accessed: 29 June 2024).
ICLEI-Local Governments for Sustainability (2018) "Multilevel Climate Action: The Path to 1.5 Degrees". Available at: https://e-lib.iclei.org/wp-content/uploads/2018/12/cCR-report-web.pdf (Accessed: 29 June 2024).
Intergovernmental Panel on Climate Change (IPCC) (2014) "Climate Change 2014: Synthesis Report. Contribution of Working Groups I, II and III to the Fifth Assessment Report of the Intergovernmental Panel on Climate Change". Available at: https://unfccc.int/documents/629614 (Accessed: 28 June 2024).
Intergovernmental Panel on Climate Change (IPCC) (2021) "Future Global Climate: Scenario-based Projections and Near-Term Information", in: *Climate Change 2021. Physical Science Basis. Working Group I Contribution to the Sixth Assessment Report of the Intergovernmental Panel on Climate Change*. Available at: https://www.ipcc.ch/report/ar6/wg1/chapter/chapter-4/#4.3 (Accessed: 6 January 2025).
Intergovernmental Panel on Climate Change (IPCC) (2022) "Summary for Policymakers", in: *Climate Change 2022: Impacts, Adaptation, and Vulnerability. Contribution of Working Group II to the Sixth Assessment Report of the Intergovernmental Panel on Climate Change*. Available at: https://www.ipcc.ch/report/ar6/wg2/chapter/summary-for-policymakers/ (Accessed: 6 January 2025).
Intergovernmental Panel on Climate Change (IPCC) (2023) "Summary for Policymakers", in: *Climate Change 2023: Synthesis Report. Contribution of Working Groups I, II and III to the Sixth Assessment Report of the Intergovernmental Panel on Climate Change*. Available at: https://www.ipcc.ch/report/ar6/syr/downloads/report/IPCC_AR6_SYR_SPM.pdf (Accessed: 28 June 2024).
Intergovernmental Panel on Climate Change (IPCC) (2024) "Planning for the Seventh Assessment Cycle: Options for the Programme of Work in the Seventh Assessment Cycle". Available at: https://apps.ipcc.ch/eventmanager/documents/83/120120240922-Doc.%204%20Rev.1%20-%20Options%20for%20Prog.%20of%20Work%20in%20seventh%20cycle.pdf (Accessed: 28 June 2024).
Japan International Cooperation Agency (JICA) (2023) "Clean City Partnership Programme (C2P2) Seminar: Urban Agenda on Climate Change, Pollution and Biodiversity Loss". Available at: https://www.jica.go.jp/english/information/seminar/2023/1529127_36691.html (Accessed: 28 June 2024).
Ministry of the Environment of Japan (MOEJ) (2021a) "Climate Action Towards Net-Zero by 2020". Available at: https://www.env.go.jp/content/000049877.pdf (Accessed: 28 June 2024).
Ministry of the Environment of Japan (MOEJ) (2021b) "Creating Sustainable, Zero-Carbon Societies Through City-to-City Collaboration". Available at: https://www.env.go.jp/earth/coop/lowcarbon-asia/english/project/data/jcm_guidbook_C2C_2021_EN.pdf (Accessed: 28 June 2024).
Ministry of the Environment of Japan (MOEJ) (2023) "G7 Roundtable Meeting on Subnational Climate Actions Summary Report". Available at: https://www.env.go.jp/content/000176032.pdf (Accessed: 29 June 2024).
Ministry of the Environment of Japan (n.d.) "Web Portal for Zero Carbon Development in Asia City-to-City Collaboration Programme". Available at: https://www.env.go.jp/earth/coop/lowcarbon-asia/english/project/ (Accessed: 29 June 2024).
UN Department of Economic and Social Affairs (UN DESA) (2018) "2018 Revision of World Urbanization Prospects". Available at: https://desapublications.un.org/publications/2018-revision-world-urbanization-prospects. (Accessed: 6 January 2025).
United Nations Framework Convention on Climate Change (UNFCCC) (2015) "Report of the Conference of the Parties on Its Twenty-First Session, Held in Paris From 30 November to 13 December 2015. Addendum. Part Two: Action Taken by the

Conference of the Parties at Its Twenty-First Session", FCCC/CP/2015/10/Add.1. Available at: https://unfccc.int/documents/9097.

United Nations Framework Convention on Climate Change (UNFCCC) (2019) "19/CMA.1 Matters Relating to Article 14 of the Paris Agreement and Paragraphs 99–101 of Decision 1/CP.21", in: *Report of the Conference of the Parties Serving as the Meeting of the Parties to the Paris Agreement on the Third Part of its First Session, Held in Katowice From 2 to 15 December 2018. Part Two: Action Taken by the Conference of the Parties Serving as the Meeting of the Parties to the Paris Agreement.* Available at: https://unfccc.int/documents/193408 (Accessed: 6 January 2025).

United Nations Framework Convention on Climate Change (UNFCCC) (2023) "Research and Systematic Observation: Draft Conclusions Proposed by the Chair", FCCC/SBSTA/2023/L.7/Add.2. Available at: https://unfccc.int/documents/630157 (Accessed: 28 June 2024).

United Nations Framework Convention on Climate Change (UNFCCC) (2024) "Report of the Conference of the Parties Serving as the Meeting of the Parties to the Paris Agreement on Its Fifth Session, Held in the United Arab Emirates From 30 November to 13 December 2023", FCCC/PA/CMA/2023/16/Add.1. Available at: https://unfccc.int/documents/637073 (Accessed: 28 June 2024).

3 The Role of Inward and Outward Foreign Direct Investment in Sustainable Smart Cities

Jesper Edman

1. Introduction

Smart cities have become an increasingly important part of Japan's energy and environmental strategy (Barrett *et al.*, 2021; DeWit, 2014; Anjali *et al.*, 2022; Deguchi, 2020). With their emphasis on efficient transport, renewable energy, smart energy grids, and advanced data collection capabilities, smart cities are often seen as a means to combine advanced technology with societal change to promote greater sustainability and overall decarbonisation (Sokołowski, 2022).

In order to achieve these sustainability goals and play their role in Japan's energy transition, smart cities require significant investment in both financial and human capital, as well as cutting-edge technologies, know-how, and societal solutions (Komninos, 2002). While national policymakers, local governments, and community activists can play an integral part in spurring such investments, they ultimately depend on cooperation and commitment from large private sector firms, as well as major universities and research centres (Komninos, 2002; Lauri, 2021). Private companies play a critical role in funding and promoting research on smart cities, while also acting as consultants and sources of best practices and knowledge transfer. Although private corporations often play an integral part in planning and realising smart city development, the public aspect of smart cities has typically received the majority of attention in extant research. A complete evaluation of Japan's smart city evolution must, however, also account for the role of private corporations that provide the technologies, capabilities, and services underlying smart city development.

In this vein, this chapter contributes to a greater understanding of the relationship between private firm's foreign direct investment (FDI) and smart city development. FDI constitutes cross-border investments into local economies, either in the form of new establishments (so-called greenfield investments), acquisitions, or local alliances (e.g. in the form of joint ventures or strategic partnerships) (Caves, 1996). FDI may encompass any of the activities in the corporate value chain, including research and development (R&D), production, sales and marketing, technical assistance, or local procurement. These

DOI: 10.4324/9781003471448-3

activities often have significant economic effects on local regions, through both direct investment and increased employment, yet they also bring non-pecuniary effects, in the form of technology transfers, learning, and network linkages to other locations (Lee and Tan, 2006; Blomström and Kokko, 1998). As a result, many policymakers view FDI as an important tool for local economic development.

In this chapter, I highlight three distinct benefits of FDI for regional smart city development. First, I discuss how *inward* FDI – i.e. investments by foreign firms into local Japanese smart city initiatives – can play a critical role in transferring capital and unique capabilities, including technology and knowledge. Second, I focus on how *outward* FDI – i.e. investments by Japanese corporations into foreign smart city initiatives – offers a means to increase learning and the development of new capabilities, which can subsequently be reapplied to the Japanese context. Finally, I discuss how the combination of inward and outward FDI creates important network linkages between local smart cities in Japan and similar municipalities in other nations. Such network linkages, I suggest, have the potential to increase the dynamism and innovative capability of smart cities, while at the same time promoting more knowledge transfer and development.

2. Inward FDI to Japanese Smart Cities

Extant research has emphasised numerous ways in which inward FDI helps promote local economic growth and development. Foreign investment often brings much-needed capital to upgrade local facilities and infrastructure, especially when the firm's activities include highly technical operations, such as manufacturing and research and development (Marin and Bell, 2006; Lee and Tan, 2006). In addition, FDI also contributes to local development through knowledge transfer and knowledge spillovers (Sjöholm and Lipsey, 2005; Blomström, 1989).

Knowledge transfer constitutes the transfer of strategy capabilities and unique sources of advantage to the local subsidiary from other units in the firm's global operations (Rugman and Verbeke, 2001). Such capabilities may include explicit advantages, in the form of technologies, patents, and other know-how, but they may also take the form of implicit knowledge, including routines, strategies, and processes (Kogut and Zander, 1992, 1993). Spillovers, in turn, are strategic knowledge and capabilities that diffuse from the foreign firm to other local competitors, for example, through processes of learning, imitation, and human capital movement (Blomström and Kokko, 1998).

While the effects of knowledge transfer and knowledge spillovers on local economic development have long been known, scholars have also highlighted their implications for shifts in local regulations and broader societal institutions (Cole *et al.*, 2006). In particular, recent work has highlighted how FDI helps in promoting the spread of sustainable business practices and acceptance for renewable energy among local communities (Doytch and Narayan,

2016; Kang *et al.*, 2021). More generally, there is an increasing recognition that the influence of foreign firms includes both investment in hard physical infrastructure and technology and broader societal practices and norms (Nyuur *et al.*, 2016). In Japan, foreign firms often struggle to make an impact, due to the highly competitive nature of Japanese firms and the unique cultural and societal traits of the local market. Nonetheless, the impact of inward FDI may be felt in terms of both knowledge transfer and physical investments.

Although many Japanese companies are highly competitive in technologies related to smart city sustainability, there are also underdeveloped sectors and market niches where foreign firms can play an integral role in further spurring smart city sustainable development. One example of this is energy-efficient and smart housing, a sector dominated by large Japanese real estate and construction firms. Building on know-how from colder European climates, for example, companies like the US firm Johnson Controls and the Swedish Gadelius KK have introduced new insulation technologies and energy-efficient technologies to boost housing production in smart cities. Working with major Japanese real estate developers like Sekisui House, these foreign players have helped spur greater recognition and interest in smart housing, which is essential for developing more sustainable buildings (Pham, 2015).

Similarly, several foreign firms have contributed knowledge and capabilities to the development of building environment monitoring systems (BEMS), with the aim of making fuel usage more energy efficient and less polluting. As an example, Johnson Controls KK, the Japanese subsidiary of a US company that specialises in heat-resistant pairing for smart buildings, partnered with Mitsui Real Estate to develop an energy optimisation system for the Kashiwanoha Smart City initiative in Chiba Prefecture. In this case, Johnson Controls' expertise was adapted to fit the Chiba smart city's needs for sustainable and disaster-resistant energy systems, specifically by working with Mitsui Real Estate to adapt and adjust its original offerings.

As the above cases imply, contributions in the form of knowledge and capability transfer often happen through collaborations and alliances with local firms, particularly in major metropolitan areas where large Japanese companies invest and have a strong presence. At times, the collaborations may also be done directly between foreign firms and local municipalities, particularly in smaller and more regional smart city initiatives. An example of such cooperation is that between the smart city of Higashi Matsuyama and Danish experts from the town of Lolland. After Higashi Matsuyama sustained significant damage in the 2011 Great Tohoku Earthquake, experts from Lolland drew on their experience and knowledge of sustainable energy already in use in Denmark to provide important knowledge and capabilities to support the rejuvenation of the town and support its smart city initiatives.

Contributions to inward investment in rural areas are not only limited to knowledge transfer but can also include investments into physical infrastructure, something that may be particularly valuable for Japanese regional smart city initiatives that lack financial resources and capabilities for implementing energy-efficient housing, sustainable energy plants, and other critical assets. While the Ministry of Economy, Trade and Industry (METI) envisions significant transformations in both urban transportation and the underlying technological infrastructure, such investments are costly and risk increasing the financial burden on local municipalities that are already suffering from falling tax income due to depopulation and ageing. Infrastructure investment by private firms can help overcome this funding gap because privately held firms tend to be more effective in their implementation (thanks to significant pre-existing information) and also because they are motivated by returns and thus have greater incentive to ensure that the project is financially feasible.

For Japanese firms, the introduction of new technologies and capabilities can offer positive learning benefits, spurring greater entrepreneurship and innovation (Barrett *et al.*, 2021; Ratten, 2017). The potentially beneficial role that FDI can play in local smart city development is recognised by Japanese policymakers. Smart cities have been linked to proposed special economic zones (Pham, 2015), and the Japan External Trade Organization (JETRO) has highlighted the smart city segment as an attractive sector to achieve the government's goal of increasing inward FDI to 80 trillion JPY by 2030 (JETRO, 2021). Moreover, in a survey of foreign firm's attitudes to the Japanese market, 13.5% indicated an interest in investing in smart cities, making it one of the top five focus areas of foreign companies. Many of the global companies with leading smart city technology – including Siemens, ABB, SAP, and Accenture – already operate in Japan.

Inward FDI thus has the potential to contribute significantly to the development of Japanese sustainable smart cities, both through knowledge transfer and investment in critical infrastructure. While these contributions can take place through collaborations with major Japanese players, they may have the most impact in more rural areas, through direct cooperation with local municipalities and townships that otherwise lack sufficient knowledge and investments. These possibilities are, however, yet to be realised due to a number of reasons, connected in particular to the dominance of Japanese companies and the limited size and investment opportunities presented by smaller smart cities. In the following section, I discuss some potential changes that can help spur changes in these areas to generate greater support for inward FDI towards smart cities.

3. Promoting Foreign Collaborations with Japanese Smart Cities

To entice greater inward FDI, local smart city initiatives can take a number of steps. First, local municipalities can increase their visibility as potential investment locations by taking part in international events, conferences, and symposia focused on sustainable smart city development. As technologies related to sustainability mature and consolidate, such gatherings can serve as important opportunities for not only directing attention at investment in

Japanese smart city opportunities, but also ensuring that the smart city initiatives themselves are in line with global sustainability standards and development. If regional Japanese smart city initiatives veer too far from global standards, putting exclusive emphasis on idiosyncratic needs and technologies, global firms may find it difficult to match their capabilities and knowledge to the realities of the Japanese context.

Second, local smart city initiatives can also increase their visibility by developing English language information materials, in both digital and physical formats. While many of Japan's leading smart city initiatives provide considerable information to prospective investors and partners, these are almost inevitably produced in Japanese. If the same material were to be available in English (the global lingua franca) or Chinese, smart cities could have a greater opportunity to attract attention from foreign investors. It would also signal their willingness to work and cooperate with foreign investors. To realise these objectives, local smart city initiatives can hire bilingual employees who can serve as consultants in global marketing efforts.

Finally, local smart city initiatives may benefit from forming collaborations and syndications with nearby communities to entice FDI. Sustainability-related infrastructure projects often require significant upfront asset investments, and the needs of a single smart city may be too small to merit attention and effort by large multinational enterprises. By banding together, regional towns, communities, and cities can increase their attractiveness to foreign firms, while at the same time offering greater benefits in the form of economies of scope and scale for the investing company and the energy and sustainability-related infrastructure project. An additional benefit of collaboration may be that communities adopt similar technologies and practices, increasing their interlinkages with each other.

In sum, inward FDI by multinational firms can serve as an opportunity for capital, technologies, and knowledge related to sustainable development. This opportunity has largely gone unrealised, however, despite support from both national policymakers and interest among foreign firms. To take full advantage of the benefits of FDI, policymakers at both the national and (especially) the regional levels must take more proactive approaches towards courting FDI.

4. Outward Alliances with Foreign Smart Cities

Sustainable development of Japanese smart cities can also benefit from – and provide benefits to – outward FDI, i.e. collaborations between foreign smart city projects and Japanese corporations and other organisational entities. The critical technologies that support smart city sustainable development – smart grids, renewable energy, artificial intelligence (AI), and mobility as a solution – are all global industries, featuring leading firms from multiple countries. From Japan, firms such as Panasonic, Hitachi, Toshiba, Mitsubishi Electric, Fujitsu, and NEC, are major competitors in their respective segments. Japanese trading houses such as Mitsui Bussan, Sumitomo Corporation, and Mitsubishi Corporation, as well as various automotive companies and real estate firms also have technology and capabilities in the sustainability-oriented elements of the global smart city industry.

Table 3.1 provides a selected overview of Japanese companies with different competencies and their roles in various smart cities. Although not exhaustive, the table indicates that many Japanese companies are focused on offering solutions related to mobility, information and communication technologies (ICT) infrastructure, and physical infrastructure. In terms of geographical location, much of the emphasis is on Southeast Asia and China, although Japanese firms are also active in Europe and North and South America, and, to a lesser extent, Africa, the Middle East, and Oceania.

When entering local markets, Japanese firms often rely on partnerships and strategic alliances with specialised local suppliers, major universities, and public municipalities. For example, in 2022, Fujitsu publicised the launch of a joint research project with Carnegie Mellon, a leading US university, focusing on digital twins and software solutions. NEC and Nomura have announced similar research collaborations with the Indian Institute of Technology and the China Institute of Information and Communication, respectively. Several companies have worked in close cooperation with local municipalities in India, Portugal, Brazil, the USA (Texas), and China to develop novel solutions and products. A number of companies have also made investments and full-scale acquisitions of firms in the USA, the UK, Singapore, China, France, Australia, Israel, and Ireland.

For Japanese companies, the most direct benefit of serving overseas smart city initiatives is that they offer growth opportunities, particularly in emerging economies such as India, Brazil, and some areas of China. By leveraging experiences and technologies developed at home, Japanese companies can become critical suppliers of various products and services for global smart city projects. Recognising this opportunity, JETRO has developed numerous case studies and reports of the various smart city opportunities available to Japanese companies in Africa, Southeast Asia, and South America, to name a few.

In order to secure access and growth, Japanese firms must have unique capabilities and assets – e.g. technologies, patents, human resources, or brands – that can provide them a competitive advantage in local markets. For Japanese companies, these advantages often derive from the unique experiences and traits of the Japanese market. First, many of the firms offer solutions linked to quality manufacturing and hardware; this includes physical ICT infrastructure, waste management, mobility infrastructure, and smart grids and other energy solutions. The notion of *monozukuri* – i.e. craftsmanship – is frequently echoed by the large multinationals in their explanation of why and how they aim to capture global market share. In this sense, Japanese offering smart city solutions differ from their global competitors, many of whom are chiefly focused on solutions, as opposed to underlying software or hardware.

Table 3.1 Japanese firms' activities in the global smart city industry

Companies	*Sector*	*Countries*	*Functional focus*	*Capabilities and expertise*	*Activities*
NEC	IT and telecommunications	India, China, UK, Spain	Development, IT telecom, mobility, software	Smart city development; consulting	Alliance, supplier contracts established
Toshiba	Electronics manufacturing	China, France, USA	Energy, overall urban development	Energy conservation, measuring, smart grid	Collaborations, alliances, acquisitions
NTT	IT and telecommunications	China, USA	Overall development, mobility, telecom	Smart city development, traffic analysis	Alliances, solution demonstrations
Mitsui Real Estate	Housing and real estate	China, Singapore, Israel, Taiwan	Residential housing, security, services	Construction and cyber-security	Local sales
Hitachi	Electronics manufacturing	China, Singapore, India, Cambodia, USA, UK, France	Consulting, mobility, IT and telecom, energy, services, waste system	Energy conservation, smart grid, autonomous driving, IoT, societal solutions	Acquisitions and greenfield establishments
Fujitsu	IT and telecommunications	USA, China, Belgium, Singapore, Indonesia	Mobility, IT, Telecom	Smart grid, software development, AI	Alliances and collaborations with universities and local partners
Nippon Koei	Consulting	Indonesia, Singapore, Cambodia, Laos, Vietnam, India	Mobility, infrastructure, general development	General, traffic management, consulting	Demonstration of solutions, direct sales

Source: Corporate press releases and annual reports, 2004 to 2023

Note: AI, artificial intelligence; IoT, Internet of Things; IT, information technology.

A second common trait among Japanese firms is their emphasis on security in smart city development. Many (albeit not all) of the firms highlight how their products ensure safer public spaces through various technologies, including public monitoring (e.g. cameras), traffic congestion analysis, digital twins, and robust infrastructures capable of withstanding natural disasters like earthquakes, typhoons, and heavy rain. NEC, for example, emphasises its ability to develop "safe and secure" smart solutions, offerings that may be of particular interest to smart cities in vulnerable or uncertain locations, or where crime is a major problem.

The expertise in safety and security presumably comes from the experience of developing domestic Japanese smart city projects and initiatives. This can serve as a unique source of competitive advantage, especially in issues related to demographic change and ageing. With one of the world's oldest populations, Japan's smart city projects often put emphasis on developing smart cities that are suitable for older citizens, a fact that is reflected in the products, services, and approaches of its leading firms. As China, South Korea, and many European countries begin to face similar demographic challenges, and demand for safe and effective smart cities grows, Japanese firms may benefit from their unique experiences in their home market. In this way, smart city development at home underpins Japanese firm's abilities to expand internationally.

In addition to leveraging their firm-specific assets and capabilities to achieve sales growth, Japanese companies also engage in overseas smart city initiatives as a means to upgrade and expand their existing knowledge base. By partnering with local universities, municipalities, and companies, Japanese multinationals gain a greater understanding of recent developments in various technological solutions, as exemplified by Toshiba's joint development of new Internet of Things (IoT) solutions with Cisco. Such collaborations also expose the Japanese to the latest smart city standards and practices, while simultaneously pushing them to upgrade their own products and offerings. In doing so, overseas FDI can further boost the competitiveness of Japanese corporations, both at home and abroad.

5. Transferring Insights from Outward FDI to Local Japanese Smart City Initiatives

Importantly, Japanese firms' experience in the global smart city sector also benefits the development of sustainable smart city initiatives in Japan itself. By transferring the knowledge acquired overseas (including both "hard" technological capabilities and "soft" insight into processes, standards, and practices) back to headquarters, firms increase the number and quality of sustainability-oriented smart city services available in Japan. Major corporations can thus take the learnings gained abroad and apply these to domestic smart city projects, thereby not only increasing their efficiency and quality but also offering new innovations and services that may have been lacking. This knowledge transfer from the global market to local Japanese initiatives is particularly crucial in areas like sustainable energy that are central to many smart city initiatives in Japan.

To ensure the effective acquisition of knowledge in global projects and their subsequent transfer back to the domestic Japanese market, large multinational firms can take several steps. First, when entering foreign markets, Japanese firms should seek to understand how to best combine their own capabilities, technology, and skills with local knowledge and expertise. Learning is not simply a question of imitating the practices and technologies of new markets; rather, it entails proactively *combining* novel insights with already existing capabilities. As an example, Fujitsu has actively sought to combine its distinct in-house skills and competencies (i.e. those developed in Japan) with the capabilities and knowledge developed by its foreign collaborators. Such active *recombinations* (as opposed to imitations or acquisitions) can help trigger the development of completely new products and services. Recombinations also increase the chance that new innovations, products, and services are accepted by the organisational members, since they are linked to their existing practices and capabilities. In this way, overseas smart city projects can serve as incubators for developing valuable new technologies, products, and solutions.

Second, in order to ensure that their overseas learning and knowledge acquisition is put to good use, Japanese firms must develop effective *reverse knowledge transfer capabilities*. Reverse knowledge transfer is the process by which the firm's subsidiaries and subunits operating in overseas locations transmit know-how back to headquarters and other subsidiaries for further dissemination and application (Ambos *et al.*, 2006). In the case of sustainable technologies and services applicable to smart cities, reverse knowledge transfer may be particularly tricky because the new knowledge developed abroad is often embedded into specific geographic needs and conditions. For example, a sustainable energy solution developed by Japanese firms to suit the regulatory framework, energy infrastructure, and climate conditions of Indonesia is unlikely to be directly transferable to the Japanese context. At the same time, many of the underlying elements of the solution – including general knowledge about how to deal with regulations or how to develop localised energy requirements – will presumably be of value to Japanese smart city initiatives. The key for firms is thus to be able to transfer the correct knowledge, in the correct way.

To ensure effective reverse transfers, firms must be able to distinguish between three kinds of knowledge gained in overseas locations: knowledge that is universally generalisable and thus readily transferable as is; knowledge which is location-specific and thus not applicable to the Japanese context; and knowledge which must first be translated and adapted to the Japanese context (Carlile, 2004; Cavusgil *et al.*, 2003; Håkanson and Nobel, 2000). The first category will typically involve physical artefacts – including infrastructure and products – whose output is unchanged between locations. For example, cameras and sensors used to monitor the efficient use of resources or renewable energy technologies are typically readily transferable. The second category – location-specific innovations – are those closely linked to local conditions and resources; in the example of Indonesia, energy systems built around local geothermal deposits might be an example of products and capabilities that are unlikely to be applicable in Japan. The third category – knowledge that is applicable after modification – is arguably the most challenging since it involves significant experimentation and exploration to understand how and whether adaptation to the Japanese context is possible. This may include positive energy houses and other sustainable housing materials whose use must be adapted to fit the Japanese culture and climate.

Achieving effective learning and knowledge acquisition from abroad rests fundamentally on human capital, i.e. the skill of employees and their specific experiences. This is particularly true for knowledge and know-how that must be adapted to fit local contexts; successful adaptation often hinges on an employee's ability to both interpret local cultures and their needs, and also to understand the possibilities and limitations of a particular technology or solution. To ensure effective reverse knowledge transfer, Japanese corporations must thus ensure that they develop employees that not only have experience from foreign smart city projects but also a good understanding of sustainable smart city initiatives within Japan. One solution to this challenge is to deploy specialised teams, consisting of product and service specialists, as well as those with community experience, who together identify and transfer overseas knowledge back to Japan. Such teams will also be instrumental in working with local smart city initiatives to tailor offerings that match the initiatives' goals and particular needs.

In sum, domestic smart city initiatives and outward FDI by large Japanese firms can develop a symbiotic relationship; the latter provides an initial testing ground to develop capabilities and solutions that are unique to Japan's sustainability goals and challenges, but which can subsequently be leveraged by Japanese firms in the global smart city market. Through their international investments and partnerships, Japanese firms can also develop knowledge and capabilities around new sustainability solutions, which can subsequently be leveraged to further effect in Japanese smart cities. For this to happen, however, the capabilities developed – both at home and abroad – must be applicable to multiple markets and also be transferable. To ensure transferability, corporations must develop organisational capabilities, especially human capital experts and global teams, which can successfully evaluate, translate, and apply knowledge about smart city developments across and within different locations.

6. FDI as a Source of Global Network Linkages

Both inward and outward FDI thus benefit smart city development through an increased flow of critical human capital, technology, and know-how. An additional benefit of this combined effect is also to establish and maintain networks between different smart cities. The local subsidiaries and subunits that are established through FDI are not stand-alone entities but rather nodes in large multinational networks that connect the various activities and operations of the firm. For both foreign and Japanese firms engaged in FDI, these networks serve as crucial avenues for exchanging information and transferring knowledge, as discussed above.

In the case of smart city-related FDI, the local subunits of multinational enterprises are likely to become deeply enmeshed and embedded among the local community and its policymakers (Meyer *et al.*, 2011). Similar to public–private partnerships, often with local governments and municipalities, smart cities are distinct from most other FDI targets in that they require integrating multiple products and services, often sourced from different firms, into a holistic package that can serve the global community well (Pan *et al.*, 2023).

As such, smart city services are not limited simply to providing a product or solution; rather, they necessitate significant involvement and relationship-building with local stakeholders in different locations. As stakeholders from local smart initiatives deepen their collaboration and relationships with corporate suppliers – whether in Japan or elsewhere – they too gain access to the firm's broader network. Consequently, the investing firm's network can serve as a conduit that connects smart cities located in different locations across the world. These networks, in turn, have various benefits for the smart city initiatives.

First and foremost, smart cities can employ their suppliers' networks to develop their own ties to similar initiatives in other locations. Such connections provide a means for smart cities to learn and exchange experiences directly, as opposed to relying on the transfer capabilities of Japanese or foreign firms. Such linkages are particularly important in the case of sustainability, where both technological solutions and political developments are developing rapidly. Moreover, the central challenge underlying the push for sustainability – i.e. climate change – is an externality, whose impact is felt far beyond the individual smart cities. In order to effectively counter the effects of climate change, both locally and globally, smart cities must work in collaboration with other initiatives and entities, both in Japan and abroad.

Second, by tapping into a global network, smart cities can play an important role in the dissemination of best practices and effective sustainability solutions. Rather than simply being recipients of knowledge, technologies, and solutions, Japanese smart cities can take initiatives to share their own unique designs, strategies and policies related to promoting sustainability. This allows smart cities to have a voice in the development of sustainability standards and approaches, which is crucial not only for the evolution of the smart city industry but also for the effectiveness of sustainability initiatives. Smart cities can also increase their bargaining power and voice vis-à-vis regulators and corporate suppliers by joining smart city collaborations and alliances, thereby gaining more power over their own development process and evolution. Such power and influence can be particularly important for smaller smart city projects, which may otherwise be dominated by national policies and corporations.

7. Discussion: Linking FDI and Japanese Smart City Development

Private corporations, both Japanese and non-Japanese, have an integral role to play in the development and growth of Japan's sustainable smart city initiatives. Private firms offer not only financial investment but also expertise and learning. At a deeper level, they can help connect regional smart city initiatives to global standards and best practices, while also undergirding direct linkages between Japanese and non-Japanese smart cities. Such networks can serve as an important source of information exchange, collaboration, and mutual development. In particular, Japanese smart cities can benefit not only from learning from foreign counterparts but also by sharing their own knowledge and expertise with other initiatives.

Figure 3.1 offers a stylised framework highlighting the linkages between Japanese smart city initiatives and foreign ditto. As the figure emphasises, foreign and domestic firms serve as primary conduits between domestic and foreign smart city initiatives, as well as the broader smart city industry and

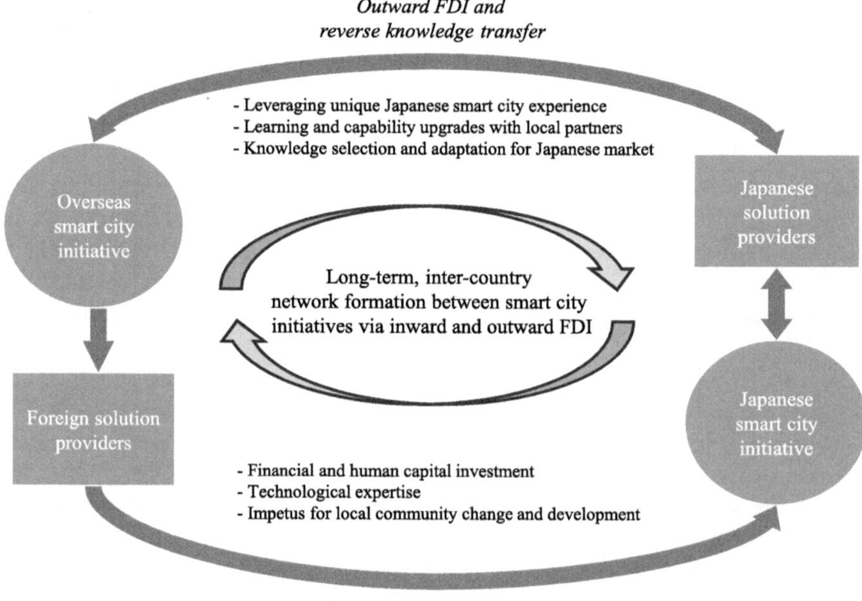

Figure 3.1 A conceptual process model of inward/outward FDI and smart city knowledge acquisition

market. As conduits, Japanese and foreign firms play somewhat differentiated roles. Foreign firms are mainly catalysts for change, bringing novel ideas and practices to Japanese smart cities, thereby supporting and furthering effective transitions towards more sustainable solutions and lifestyles.

Japanese companies, in turn, play a more long-term stabilising role, primarily because they are more embedded in the home market and draw much of their own sources of competitive advantage from developments in the Japanese smart city space. While Japanese companies contribute knowledge and know-how gleaned overseas, they are typically better equipped to translate and localise such knowledge to fit the needs of local smart city initiatives. Moreover, given their linkages to local stakeholders – including communities, municipalities, and domestic firms – Japanese firms' innovations and novelties will presumably be more incremental in nature, often designed to align with existing practices and behaviours, as opposed to challenging or redefining them. By each playing a distinct role in the evolution of Japanese smart cities – either as supporters of radical change and innovation (e.g. foreign firms) or sources of stability and incremental change (Japanese firms) – the two different sources of FDI can balance each other, creating a fruitful dynamic for local development.

The potential positive benefits envisioned in the framework depend on several factors. First, the scale of smart city initiatives will likely be critical to enticing attention and investment from private firms. Smaller smart city initiatives, or those whose scope is limited to particular functions rather than overall urban transportation, may struggle to generate interest from major companies. This is especially true in relation to foreign firms, who face the additional hurdle of overcoming cultural or regulatory differences in their local investment, but it also applies to Japanese companies. Smart city solutions are dominated by large firms, who often rely on economies of scale and scope to reap returns on investment; without significant size, these firms may refrain from investing in local initiatives. As noted above, smart city projects can increase scale by forming collaborations and syndications, thereby increasing the overall market size and potential return for investors. In addition, smart cities may also seek to encourage the establishment of other firms within similar industries, thus creating the basis for innovation hubs and entrepreneurial clusters around smart city technologies and solutions.

A second challenge already alluded to above is the potential regulatory or cultural barriers that corporate investors may face, particularly non-Japanese firms. These may range from local suspicion and opposition to new technologies and solutions (e.g. those related to renewable energy or new mobility systems), as well as restrictions on data collection and privacy, and specific technical requirements (e.g. for local energy grids, telecommunications). While such barriers and forms of opposition are reality for all FDI, they may be particularly salient in relation to smart cities, since these encompass not only technical solutions but also broader changes in the overall structure and functioning of society and its many institutions.

Overcoming local barriers and opposition requires proactive efforts from regulators, both at the municipal and national levels. The latter – including JETRO, METI, and other ministries – should oversee rules and regulations on sustainable investment, and also open dialogue with potential partners to understand where current barriers arise and how they can be resolved and overcome. In turn, local municipalities and community organisers must put efforts towards promoting smart city investments and initiatives among their citizenry, for example, through marketing campaigns and information sessions. They must also ensure broad support and buy-in from important local stakeholders and organisations to avoid conflict and opposition once smart city changes begin to be implemented.

Finally, in order to sustain the flow of knowledge and capabilities across borders, and to avoid any investments into smart initiatives from simply becoming one-off transactions, smart city initiatives must put effort into building networks with other like-minded projects, not only in Japan but in other countries and regions as well. The capacity for smart cities to achieve the stated sustainability and economic growth goals depends on access to resources, technologies, ideas, and practices, many of which can be accessed via global networks. Such networks do not arise independently; rather, they must be nurtured and actively cultivated. Large multinational investors can provide the initial impetus and opening for developing ties to other smart cities, but over time, local municipal leaders must themselves manage these connections. Such collaborations and networks are crucial to ensure that smart cities are able to reach their goals in the long-term shift to more sustainable energy sources and operations.

Strengthening ties with other smart cities and encouraging knowledge exchange can be done in several ways. First, municipal leaders and representatives of local smart city projects should take part in global meetings and conferences, such as the Global Smart City Alliance and the World Smart Cities Forum. Such groupings and their annual gatherings provide opportunities to both deepen existing ties and establish new ones, while at the same time exposing leaders to the latest developments in smart cities. International gatherings and conferences also provide opportunities for representatives of Japanese smart city communities to spread information and knowledge based on their own experience, thereby increasing touch points with the global industry. Of particular value is if smart city representatives can take on leadership positions in global organisations, as this would provide both greater visibility and a vantage point to observe the field. National regulators and industries should support such efforts by providing travel stipends and other forms of support to delegates.

On the community level, opportunities to strengthen ties should be provided through regular exchanges and interactions with other smart city initiatives. Such exchanges can happen at multiple levels, including that of secondary schools and universities, small- and medium-sized enterprise, local government, or other vital functions. Ties can also be strengthened through the formation of global teams that bring together representatives from multiple smart city initiatives with shared challenges, experiences, or concerns. Consider, for example, if smart cities are facing water shortages, they can establish joint teams that can address potential solutions, which in turn can be tailored to each location's needs. By joining such teams, representatives from Japanese smart cities have an opportunity to both learn and influence the evolution of smart city solutions and technologies.

8. Conclusion

Smart cities are an important element in the future sustainability and rejuvenation of Japanese regional economies. While much of the onus on development lies on policymakers, both at the national and local levels, private firms must also be leveraged as key partners and assets in smart city development. While local Japanese firms can often find effective solutions, it is vital that Japanese smart cities are able to access global technologies, ideas, and practices; one way to identify and access these is through increased inward and outward FDI in the smart city industry. FDI can provide both a means of learning and technology transfer, as well as a helpful yardstick to ensure alignment with global standards and practices. Perhaps more importantly, it offers an opportunity to leverage smart city capabilities in new locations, while at the same ensuring a long-term focus on development and growth at home.

Multinational corporations – in the form of both global Japanese companies such as NEC, Hitachi, Fujitsu, and Toshiba – as well as foreign investors – including Siemens, Microsoft, Cisco, and Johnson Controls – can play an important role in connecting smart city initiatives with both the global knowledge base in general and other smart initiatives in particular. By working proactively with and towards these private companies, local Japanese smart city initiatives can create their own global networks and thus further support and ensure their long-term development.

References

Ambos, T.C., Ambos, B. and Schlegelmilch, B.B. (2006) "Learning From Foreign Subsidiaries: An Empirical Investigation of Headquarters' Benefits From Reverse Knowledge Transfers", *International Business Review*, 15, pp. 294–312.

Anjali, S.K., Aki, S. and Kenji, I. (2022) "Tokyo Smart Global Megacity: Smart Sustainable Energy Solutions", in T.M. Vinod Kumar (ed) *Smart Global Megacities: Collaborative Research: Tokyo, Mumbai, New York, Hong Kong-Shenzhen, Kolkata*. Singapore: Springer, pp. 191–218.

Barrett, B.F., Dewit, A. and Yarime, M. (2021) "Japanese Smart Cities and Communities: Integrating Technological and Institutional Innovation for Society 5.0", in H. Min Kim, S. Sabri and A. Kent (eds) *Smart Cities for Technological and Social Innovation*. London: Academic Press, pp. 73–94.

Blomström, M. (1989) *Foreign Investment and Spillovers*. London: Routledge.

Blomström, M. and Kokko, A. (1998) "Multinational Corporations and Spillovers", *Journal of Economic Surveys*, 12, pp. 247–277.

Carlile, P.R. (2004) "Transferring, Translating, and Transforming: An Integrative Framework for Managing Knowledge Across Boundaries", *Organization Science*, 15, pp. 555–568.
Caves, R. (1996) *Multinational Enterprise and Economic Analysis*. Cambridge: Cambridge University Press.
Cavusgil, S.T., Calantone, R.J. and Zhao, Y. (2003) "Tacit Knowledge Transfer and Firm Innovation Capability", *Journal of Business & Industrial Marketing*, 18(1), pp. 6–21.
Cole, M.A., Elliott, R.J. and Fredriksson, P.G. (2006) "Endogenous Pollution Havens: Does FDI Influence Environmental Regulations?", *Scandinavian Journal of Economics*, 108, pp. 157–178.
Deguchi, A. (2020) "From Smart City to Society 5.0", in Hitachi-UTokyo Laboratory (ed) *Society 5.0: A People-Centric Super-Smart Society*. Singapore: Springer, pp. 43–65.
DeWit, A. (2014) "Japan's Resilient, Decarbonizing and Democratic Smart Communities", *Asia-Pacific Journal-Japan Focus*, 12, pp. 1–16.
Doytch, N. and Narayan, S. (2016) "Does FDI Influence Renewable Energy Consumption? An Analysis of Sectoral FDI Impact on Renewable and Non-Renewable Industrial Energy Consumption", *Energy Economics*, 54, pp. 291–301.
Håkanson, L. and Nobel, R. (2000) "Technology Characteristics and Reverse Knowledge Transfer", *Management International Review*, 40, pp. 29–48.
JETRO (2021) "JETRO Invest Japan Report 2021". Available at: https://www.jetro.go.jp/en/invest/investment_environment/ijre/report2021/ch2/sec1.html (Accessed: 30 April 2024).
Kang, X., Khan, F.U., Ullah, R., Arif, M., Ur Rehman, S. and Ullah, F. (2021) "Does Foreign Direct Investment Influence Renewable Energy Consumption? Empirical Evidence From South Asian Countries", *Energies*, 14, p. 3470.
Kogut, B. and Zander, U. (1992) "Knowledge of the Firm, Combinative Capabilities, and the Replication of Technology", *Organization Science*, 3, pp. 383–397.
Kogut, B. and Zander, U. (1993) "Knowledge of the Firm and the Evolutionary Theory of the Multinational Corporation", *Journal of International Business Studies*, 24, pp. 625–645.
Komninos, N. (2002) *Intelligent Cities: Innovation, Knowledge Systems and Digital Spaces*. London: Spon Press.
Lauri, C. (2021) "Expert Knowledge and Smart City Administration", *European Review of Digital Administration & Law*, 2(1), pp. 57–76.
Lee, H.H. and Tan, H.B. (2006) "Technology Transfer, FDI and Economic Growth in the ASEAN Region", *Journal of the Asia Pacific Economy*, 11, pp. 394–410.
Marin, A. and Bell, M. (2006) "Technology Spillovers From Foreign Direct Investment (FDI): The Active Role of MNC Subsidiaries in Argentina in the 1990s", *Journal of Development Studies*, 42, pp. 678–697.
Meyer, K.E., Mudambi, R. and Narula, R. (2011) "Multinational Enterprises and Local Contexts: The Opportunities and Challenges of Multiple Embeddedness", *Journal of Management Studies*, 48, pp. 235–252.
Nyuur, R.B., Ofori, D.F. and Debrah, Y.A. (2016) "The Impact of FDI Inflow on Domestic Firms' Uptake of CSR Activities: The Moderating Effects of Host Institutions", *Thunderbird International Business Review*, 58, pp. 147–159.
Pan, A., Zhang, W. and Zhong, Z. (2023) "How Does FDI Affect Cities' Low-Carbon Innovation? The Moderation Effect of Smart City Development", *Emerging Markets Finance and Trade*, 59, pp. 1247–1261.
Pham, C. (2015) *Tokyo Smart City Development in Perspective of 2020 Olympics: Opportunities for EU-Japan Business Cooperation and Development*. Tokyo: EU-Japan Centre for Industrial Cooperation. Available at: https://cdnw8.eu-japan.eu/sites/default/files/publications/docs/smart2020tokyo_final.pdf (Accessed: 30 April 2024).
Ratten, V. (2017) *Entrepreneurship, Innovation and Smart Cities*. London: Routledge.

Rugman, A.M. and Verbeke, A. (2001) "Subsidiary-Specific Advantages in Multinational Enterprises", *Strategic Management Journal*, 22, pp. 237–250.
Sjöholm, F. and Lipsey, R.E. (2005) "The Impact of Inward FDI on Host Countries: Why Such Different Answers?", in T.H. Moran, E.M. Graham and M. Blomström (eds) *Does Foreign Direct Investment Promote Development?* Washington, DC: Institute for International Economics: Center for Global Development.
Sokołowski, M.M. (2022) *Energy Transition of the Electricity Sectors in the European Union and Japan: Regulatory Models and Legislative Solutions*. Cham: Palgrave Macmillan.

4 "Smart" Cities and Dumb Solutions

The Risks of Technology-Reliant Solutions for Decarbonisation in Japan

Jordan T. Carlson and Gregory Trencher

1. Introduction

For nearly two decades, the Japanese government has implemented various programmes to pursue the development of smart cities (Barrett *et al.*, 2021). By providing funding for the demonstration or implementation of new technologies and by building platforms for otherwise competing firms to cooperate and accelerate innovation by joint learning (Nyberg and Yarime, 2017), Japan is now home to dozens of smart city developments in both urban and semi-rural areas. These have sought to advance environmental sustainability, innovation-driven economic development, digitalisation, and resilience against natural disasters such as earthquakes (Sakuma *et al.*, 2021). They have also attempted to tackle societal challenges, such as population ageing and health (Trencher and Karvonen, 2017; Trencher, 2019). Many of Japan's smart city projects are called "smart communities". These interpret the concept of "community" broadly (DeWit, 2018), using smart energy systems to link residential districts, public facilities, factories, commercial hubs, and other zones of sufficient demand. Granier and Kudo (2016) argue that the aspiration of Japan's smart communities was to go beyond implementing novel technologies to make industrial, commercial, business, and citizen behaviour itself "smart". The works by Sokołowski (2021) and Kusanagi (2023) have linked these efforts to conceptions of "energy communities" within Japan. Yet they suggest that Japanese smart communities are still more aligned with the centralised, technocratic visions of smart cities critiqued in other countries, failing to realise goals related to behavioural change. In more recent scholarship, this evaluation has been both supported (Sakuma *et al.*, 2021) and challenged (Barrett *et al.*, 2021). Yet both sides agree that energy technology implementation is a central focus of Japan's smart cities.

Due to this focus on technological development, it has been argued that Japanese smart city policies are more an example of industrial policy rather than energy transition or climate policy (DeWit, 2014, 2018; Nyberg and Yarime, 2017). This implication is frequently visible in the stakeholder groups that spearhead smart city development. Although it is often municipalities

DOI: 10.4324/9781003471448-4

that claim ownership or assume responsibility for project management, large technology and engineering firms typically play the biggest role in project conception and implementation. Despite the myriad complementary and conflicting objectives that underpin Japan's efforts to build smart cities, a key feature running across most is the ambition to reduce carbon emissions (Sakuma *et al.*, 2021).

The extant literature on smart cities abounds with critical accounts of smart city developments around the world where the goals of corporations or economic development have taken priority over environmental or social aims (Martin *et al.*, 2018). Cugurullo (2019) and Kitchin *et al.* (2019) highlight that smart cities are subject to many competing interests in their planning and development, emerging from different histories and contexts within each city that pursue "smart" policies. These differences are often flattened when economic and political interests begin competing with proposed social and environmental goals, with economic interests often winning out in many cities (Luque-Ayala and Marvin, 2015; Clement and Crutzen, 2021). Many works, such as that by Ahvenniemi *et al.* (2017), have also highlighted the reality that smart city policies are not, by definition, sustainable city policies – though are often branded as such. Thus, any potential conflict between the stated aim of achieving decarbonisation found in many Japanese smart cities plans and the reality of implemented policies and technologies is of interest.

In this chapter, we approach this conflict by introducing the concept of *decarbonisation risks*. We define such risks as any vulnerability, uncertainty, disruption, or hazard presented by a proposed technology, resource, or end use to the goal of reducing greenhouse gas (GHG) emissions (Carlson and Trencher, 2024). This definition draws on common conceptions of risk (Gurtu and Johny, 2021), wherein the concern is the vulnerability, uncertainty, or disruption caused by an action or technology to a particular person, group, or interest. We propose the use of this understanding of decarbonisation risks to assess the successes, failures, and trade-offs in policy and technology choices as governments, corporations, and citizens pursue decarbonisation of their shared societies.

From this perspective, this chapter aims to identify and discuss the implications of decarbonisation risks present in exemplar projects within Japanese smart cities. We consider three areas that Japan has emphasised in its smart cities to date: mobility, built environment, and energy management. We highlight the emphasis on implementing novel technologies developed by the Japanese industrial partners in each of these efforts, often occurring to the exclusion of simpler, "dumb" solutions that may achieve greater decarbonisation progress. Situating these discussions within the broader literature, we conclude by making recommendations for win-win decarbonisation policies within existing and future Japanese smart city projects.

This chapter makes an important contribution to the literature. Japan's smart city experiments offer crucial opportunities to reconsider the risks embodied in various approaches to pursuing decarbonisation, many of which on the surface appear "smart". Critically examining the risks involved in these approaches is essential, since all countries around the world must rapidly increase their rate of decarbonisation if we are to succeed in keeping post-industrial global heating to within 1.5° C or 2.0° C as agreed under the Paris Climate Agreement (International Energy Agency, 2023). The world's rapidly shrinking climate budget means that there is no room for decarbonisation solutions to fail. Accordingly, we address the lack of longitudinal studies thus far that has hampered a wider understanding of Japan's successes and challenges in achieving decarbonisation outcomes, which are often assumed or taken for granted (DeWit, 2018; Yarime, 2020; Sakuma *et al.*, 2021; Sokołowski, 2022). Our work also seeks to shine a light on an under-considered aspect of smart city planning and identify promising areas for future research and the development of countermeasures with policy and changed practices.

2. Mobility

Many smart city initiatives in Japan have included mobility projects. These include deployment of electric vehicle infrastructure, demonstrations of autonomous passenger vehicles, robot delivery vehicles, light-rail transit projects and bike or car sharing programmes. Efforts to support the adoption of electric vehicles have occurred notably in Kashiwanoha (Hertz, 2023), Fujisawa Sustainable Smart Town (Sakurai and Kokuryo, 2018), and Yokohama (City of Yokohama, n.d.), along with Kitakyushu's Hydrogen Town project, which also supports the deployment of fuel cell vehicles (FCV) (City of Kitakyushu, 2022). Kashiwanoha has also established itself as a leading experimenter with autonomous vehicles, notably by integrating self-driving technologies on a commuter bus route between Kashiwanoha Station and Tokyo University's nearby campus (Kashiwa City *et al.*, 2019; Shimono, 2020). Fujisawa's smart community has also experimented with autonomous mobility in the form of parcel delivery robots (Hornyak, 2022). Meanwhile, the Woven City by Toyota – currently being built from scratch at the site of the company's former factory in Susono, Shizuoka Prefecture – will act as a showcase for Toyota's autonomous vehicles and mobility solutions, including the e-Palette (Bjarke Ingels Group, 2020; Akabane, 2024). According to the plans, Toyota's e-Palette will be used in Woven City for transport and delivery services, as well as for mobile retail, food, medical clinics, hotels, and workspaces. This corporate vision of a future city where stores, hotels, and restaurants come to the consumer on demand paints a picture of great convenience, but not necessarily of more efficient, climate-conscious energy use.

Some smart city projects across Japan have made notable efforts to entice modal shifts to more sustainable forms of mobility. These include the city of Toyama's investments in light rail (Ito and Kawazoe, 2022), part of the local government's pursuit of the "compact city" ideal, as well as the ongoing efforts in Osaka to ensure walkability as a key feature of the city's neighbourhoods

(Kato and Kanki, 2020). They also include e-bike share programmes, which have featured in several projects across the country (Gazoo, 2020; Okubo *et al.*, 2022), and other efforts to maintain or expand the share of non-car commuting within Japan. Such efforts make perfect sense in Japan, which already has a high share of non-automobile commuting. For example, in Tokyo, 36% of personal transport occurs via public transit (trains 33%, buses 3%), a further 23% by walking, and 13% by bicycle. Conversely, only 28% of personal transport in Tokyo relies on private motor vehicles (Itokawa, 2020). In contrast, private cars account for 36% of personal transport in London, UK, and for 81% in Houston, Texas, USA (Itokawa, 2020). This puts Tokyo significantly ahead of post-industrial country peer cities in low-carbon personal transport in urban spaces, which is a key advantage for decarbonisation planning (and a feature shared with most other major Japanese metropoles).

Yet vehicles remain a common feature of many Japanese smart cities. This continued reliance on personal vehicles carries significant decarbonisation risks. Recent work has highlighted that zero-emission vehicles like electric vehicles (EVs) and FCVs embody significant upfront investments of carbon and energy, which requires long service lifetimes before carbon savings are realised compared to replaced gasoline vehicles (Ren *et al.*, 2023). There is also a carbon risk posed by public misconceptions of cleaner technologies, depending on which are implemented (Carlson and Trencher, 2024). The rebound effect has been widely studied in the case of hybrid vehicles (de Haan *et al.*, 2007; Muratori and Rizzoni, 2016; de Blas *et al.*, 2020), though the concerns for grid stability due to shifting electricity demand peaks seem of greater risk for EVs. At the same time, the energy demand of EVs and FCVs may cause lock-in of existing fossil-fuel infrastructure. This lock-in can occur both domestically, with polluting power plants, and overseas in hydrogen production facilities, as more than 98% of the world's hydrogen comes from fossil fuels (International Energy Agency [IEA], 2021).

The carbon benefits of autonomous vehicles are also widely challenged. Concerns have been raised that their door-to-door convenience may erode demand for public transport or reduce the attractiveness or safety of cycling and walking (Ward *et al.*, 2021). Some worry that increases in vehicle miles travelled may result from turning travel time into "idle time", reducing drivers' urgency to reach destinations due to the ability to handle other tasks while riding in autonomous vehicles (Wadud *et al.*, 2016; Moriarty and Wang, 2017). In the case of Rome, Patella *et al.* (2019) showed that if the entire vehicle fleet were electrified, autonomous vehicles would lead to 6% greater lifecycle emissions than non-autonomous ones due to increased construction, maintenance, and end-of-life emissions relating to vehicle sensors, networking, and other material-intensive equipment. Ward *et al.* (2021) have shown that autonomous vehicles lead to 20% increases in fuel consumption and GHG emissions due to "deadheading" – travel occurring without passengers. They further found that any displacement of public transit or active transport methods by autonomous vehicles worsens this effect by up to a factor of three, to 60%. These effects could be lessened through vehicle electrification, but other issues highlighted regarding road maintenance and vehicle demand remain, in line with Patella *et al.* Within Japan, Zhou *et al.* (2023) investigated the environmental performance of shared autonomous vehicles displacing private vehicles for elderly citizens in two towns, finding emissions reductions only with full electrification and high vehicle use frequency. While autonomous vehicles may play some part in future decarbonisation efforts, it is difficult to regard them as a first solution when walking, cycling, and public transit are known to be better enablers of reduced emissions.

It is worth noting here that much of this emphasis on demonstration of personal mobility technologies may arise from the proponents of Japan's smart cities, which are dominated by large firms and technology vendors (Nyberg and Yarime, 2017). Panasonic is the premier sponsor of the Fujisawa Sustainable Smart Town project and emphasised their support for EV chargers and home battery-photovoltaic (PV) systems in their own publications about the project (Fujisawa Sustainable Smart Town, 2023). Toyota broke ground on its Woven City project in 2021, and corporate materials reference the roles of EVs, FCVs, and autonomous "E-Pallets" in the city vision documents (Bjarke Ingels Group, 2020). Nissan has also been a key supporter of EV projects in Japanese smart cities, with deployments in Aizuwakamatsu, Namie, Futaba, and Minamisoma in Fukushima Prefecture (Nissan, 2021; Autovista24, 2022). Each of these private companies, ultimately, has an interest in developing, demonstrating, and selling their products. This interest is perfectly rational, but it may not be well aligned with the goal of maximising decarbonisation efforts through existing and proven technologies – much less, in the case of car companies and their partners, reducing vehicle sales in favour of public transit, cycling, and walking. These simpler solutions can lead to significant carbon reductions, but may not be "smart" enough for smart cities. The emphasis on personal vehicles also plays into a larger issue with some Japanese smart cities: a concerning emphasis on new-build, single-family housing developments.

3. Built Environment

A common feature in most of Japan's smart cities is that developers build them from scratch (Hornyak, 2022). This has led to greenfield developments alongside redevelopment projects where old industrial sites have been replaced with new mixed-use residential and commercial districts (Sunikka-Blank and Kiyono, 2021). Some greenfield projects have been unavoidable, especially in the Tohuku region post-2011 during reconstruction efforts that unfolded in tsunami-affected areas (e.g. Higashi-Matsushima). However, although many new-build smart cities have emerged as redevelopment projects, very few have focused on refurbishment or renovation of the existing urban building stock. Most developments have instead concentrated on building new "smart" homes, with energy innovations focused on solar PV, fuel cells, batteries, and integration into grid management systems (Chen *et al.*, 2020; Del Rio *et al.*,

2021; Li *et al.*, 2022). Although such approaches can lead to emissions reductions and increased resilience to disruption from disaster, there are significant questions about the long-term carbon benefits of Japan's approach.

Scholars and industry around the world have studied the trade-offs involved in refurbishing old buildings as opposed to building new ones. There is broad agreement that there are significant carbon emissions savings – and environmental co-benefits from avoiding waste of other materials – in refurbishment and renovation projects. In the United Kingdom, across housing types, refurbishments of existing homes to modern standards can avoid 30% more emissions than equivalent replacement (Schwartz *et al.*, 2022). Berrill *et al.* (2022) have shown the importance of renovating the existing housing stock in the United States rather than relying on replacements to achieve the nation's decarbonisation goals. Within Japan, some companies (e.g. Renoveru, 2019) claim similar results, with renovation for energy efficiency and sufficiency being a much lower emission path.

At various points over the past decade, refurbishing existing buildings has gained attention as a more affordable and environmentally preferable choice for construction. This has given rise to so-called *rinobēshon*[1] ("R-Innovation") – "renovation" that is innovative due to how rare it has been in Japan over the past decades (Hornyak, 2022; Nagata, 2023). This paradigm breaks with the dominant demolish and rebuild mentality that has characterised Japan's approach to urban development since the 1950s. It also offers an attractive solution to mitigating the rising costs of construction following the weakening of JPY since early 2021. Renovating can further help mitigate the chronic labour shortages slowing construction projects in Japan due to its shrinking workforce caused by population ageing (YKKAP, n.d.). Some firms have studied the climate and environmental benefits of their renovation approaches in Japan, with findings in line with international work: reductions of greater than 70% in emissions and 90% in material waste compared to new-build (Renoveru, 2019). But private firms are not the only actors pushing for change in the built environment within Japan.

In spring 2023, the Japanese government began enforcing new rules (i.e. building codes) on energy efficiency for all new buildings. These rules, part of the Act on Rationalizing Energy Use and Shifting to Non-fossil Energy (METI, 2023), call for all new-build construction for non-residential purposes to meet the national zero emission building (ZEB) standards from 2024, and for all new residential buildings to meet the national zero emission home (ZEH) standards from 2030 (ECCJ, 2022). This law also includes support programmes for upgrading existing buildings in these plans, but such programmes are incentives rather than mandates. It should also be noted that while the mandatory nature of new-build efficiency standards is positive, the existing voluntary industry standard HEAT20 (set forth by the Society of "Hyper-Enhanced insulation and Advanced Technology houses for the next 20 years", an industry group deriving its acronym as-capitalised here) has more stringent rules for qualification than this government baseline (HEAT20, 2020). This suggests that the new standards set could be made more ambitious. However, these industry standards and recent corporate discussions of "R-Innovation" have yet to be reflected in existing smart city projects in Japan.

Japan's preference for new buildings can be traced to multiple historical trends. Modern Japanese housing policy has largely been to lease or sell land to developers with few restrictions on what they can build on it (Sunikka-Blank and Kiyono, 2021). This *laissez-faire* approach has ensured a steady supply of housing to the nation, including through its period of rapid economic growth in the latter part of the 20th century. However, this approach created perverse incentives for developers: when revenue depends on selling new houses, there is little reason to build homes designed to last or to renovate existing homes. As a result, Japanese housing has frequently been considered lower in build quality than in many peer countries (Wuyts *et al.*, 2019; Sunikka-Blank and Kiyono, 2021). Public attitudes to housing were also shaped by these decisions, with most homebuyers having a strong preference for new-build homes (whether detached or apartment) rather than the second-hand ones. This preference is reinforced by the tendency for homes built since the 1980s to fall quickly into poor condition, thus losing their comfort and value (Berg, 2017). This is reflected in the average age of homes in Japan: 25 years, contrasting sharply with American and British averages of 61 and 100, respectively (Wuyts *et al.*, 2019). Exacerbating this trend is the reluctance of most Japanese homeowners to invest in refurbishments to extend the life or comfort of their homes (Wuyts *et al.*, 2019; Nagata, 2023).

The development corporations responsible for this historical pattern and the modern material culture of the Japanese housing market have also been key players in the smart city paradigm explored so far. Okubo *et al.* (2022) note that six of nine smart cities they reviewed were led by house building or real estate companies. This mirrors the prominence of public–private partnerships within Japan's smart city mobility projects. The reliance on incumbent private actors to drive smart city developments means that emission reduction strategies that run counter to these established interests may be deprioritised, thus compromising the potential decarbonisation efforts of Japan's smart cities.

That said, Japan's preference for new construction has not been without environmental benefits (Hornyak, 2022). Okubo *et al.* (2022) highlight that all smart cities they reviewed ensured high heat insulation standards, home-shading architecture techniques for the creation of cool air, and the integration of urban greening into city plans. These approaches can significantly reduce energy use in the city by reducing summer heat island effects and helping maintain indoor temperatures in winter, leading to less air conditioner and heater use year-round. Similarly, urban design strategies that pursue a "compact city" – one in which people can easily walk, cycle, or take transit to work – can contribute to these decarbonisation gains by enhancing the environmental performance of Japan's already ubiquitous mixed-use neighbourhoods centred on rail and bus stations. Toyama's light rail projects and conscious efforts to contain urban sprawl, discussed earlier, provide a perfect example of such practice (Ito and Kawazoe, 2022). Despite such successes with integrated urban planning and non-automobile transport in some compact cities, there remains a glaring deficiency in Japan of smart cities that pursue the revitalisation and decarbonisation of existing buildings as a major objective.

4. Energy Management

Home and community energy management have been a key focus of Japanese smart city projects nationwide. Japan has particularly emphasised digital monitoring and control of household energy through Home Energy Management Systems (HEMS) and their community and building equivalents (Community Energy Management Systems [CEMS]; Building Energy Management Systems [BEMS]) (DeWit, 2018; Deguchi, 2020; Li *et al.*, 2020; Barrett *et al.*, 2021; Okubo *et al.*, 2022). These systems typically entail the integration of panels or apps showing residents' real-time energy consumption data, enabled by sensors. HEMS also allows grid operators to "smartly" manage a home's solar panel and battery output to the grid, or allow residents to schedule their energy-intensive appliances (e.g. water heaters, washers) to run during off-peak hours. At least, these are the proposed advantages put forth by advocates for these systems, including Fujitsu's work in Aizuwakamatsu, Panasonic's Fujisawa Smart Sustainable Town, Mitsui's work in Kashiwanoha, and Toyota's Woven City (Tada *et al.*, 2014; Sakurai and Kokuryo, 2018; Kashiwa City *et al.*, 2019; Bjarke Ingels Group, 2020).

However, reviews of HEMS use in practice find more varying results for decarbonisation than advocates might prefer. Iwafune *et al.* (2017) conducted a national survey of 2,000 Japanese respondents in HEMS-equipped homes, finding that electricity savings were modest, limited to between 3% and 7% depending on season. Kc *et al.* (2018) conducted a comparative study of home air conditioning and heating use by residents in condominiums in Shinagawa with HEMS installed. Comparing prior studies of non-HEMS buildings, they found that resident behaviour was not different between the two heating and cooling systems: if it was cold, residents would put on a sweater or other layers, while if it were hot, they would open a window. While the HEMS building residents had lower energy use, the authors attribute this to energy-efficient building design choices, not differences in resident behaviour. Similarly, Li *et al.* (2022) found that while residents of HEMS-equipped homes in Kitakyushu used less energy than those without, 56% did not check their HEMS data reports regularly, and 83% rarely checked the community report equivalents. It is therefore worth asking if similar or better emissions reductions could be achieved with less sophisticated solutions that rely less on expensive technologies with modest returns.

However, beyond these concerns, Japan is at the leading edge of a profound demographic transformation occurring across the globe: population ageing and shrinking (commonly called an "ageing society"). Ageing has been a major topic of discussion in Japanese smart city discourses (Trencher and Karvonen, 2017; Deguchi, 2020), and it is relevant to decarbonisation for several reasons. Older people are more ingrained in their habits, face an increasing digital divide, and are less likely to adopt new technologies or ways of doing things, visible today in the adoption rates of smartphones and new mobility solutions (Harvey *et al.*, 2019; Choudrie *et al.*, 2020). Yagita and Iwafune (2021) highlight these potential challenges to decarbonisation efforts through interviews with elderly people in Japan regarding the purchase of more efficient appliances and modal shifts in heating and cooling their homes. They found that elderly people are less inclined to trust a scientist, government official, or corporate representative suggesting energy-focused lifestyle changes through replacing appliances or behaviour adjustments. Instead, their participants were more inclined to trust salespeople or technicians who have helped them manage problems with their appliances in the past. One solution to this challenge could be a people-centred technology diffusion strategy where elderly residents are engaged to move to lower carbon and more efficient technologies through cooperation with local businesses rather than corporate or government programmes (Yagita and Iwafune, 2021). Yet these solutions must be legible to end users, and the limited technological literacy of the eldest in an ageing society will prove a barrier to decarbonisation strategies that over-emphasise "smart" solutions over effective reductions.

5. Discussion and Solutions

The trends in Japanese smart city projects outlined in this chapter raise several questions. Are the novel, and often expensive, technologies modelled in predominating smart cities worth the decarbonisation risks imposed? Are the solutions implemented in Japanese smart cities reflecting the interests of certain firms rather than pursuing the most common sense and low-risk paths to emissions reductions? Has Japan overlooked "dumb" but more certain solutions to decarbonisation in prioritising smart technologies? Here, we will consider these questions in turn, and explore some possible solutions for future Japanese – and global – smart city projects that wish to pursue decarbonisation.

5.1 Smart Cities and Decarbonisation Risks

> *Are the novel, and often expensive, technologies modelled in existing smart cities worth the decarbonisation risks imposed?*

There are clearly significant decarbonisation risks imposed by Japan's smart city strategies to-date. Locking in carbon-intensive infrastructure and technologies, or choosing slower paths to societal decarbonisation than peer countries, may significantly impact the ways of life available to people in Japan. Such risks emerge in each of the three areas examined thus far: mobility, built environment, and energy management.

Continued use of automobiles as the major means of transport for the public comes with the risk of continued emissions from manufacturing, road maintenance, and (for FCVs) concerns around the means of hydrogen production. Even in the case of full electrification, the energy and emissions intensity of private or shared automobiles will always be vastly higher than that of transit, cycling, and walking. Smart city strategies that emphasise EVs over public and active transit will therefore face greater challenges realising total system decarbonisation goals. These challenges will be exacerbated by increased road maintenance frequency and urban heat island effects resulting from the larger paved-surface area inherent to automobile-friendly city designs, with the latter likely to contribute to increased air-conditioner use and energy demand.

A further area in some of Japan's smart cities where prevailing approaches to urban planning risk exacerbating decarbonisation risks is the ongoing rollout of new detached homes and car-based suburban commuter lifestyles. Developing new homes from scratch results in significantly more carbon emissions than refurbishing existing buildings. Though a few corporate advocates and the media have promoted the revitalisation of existing buildings through the "R-innovation" concept, this approach to construction is yet to mainstream in Japan's housing sector, where a preference for new construction continues. But suburban car-dependent developments innately take up more space, consume more energy, and encourage ownership of private automobiles. This paradigm must be critically assessed if put forward as decarbonisation solutions in smart cities. Though Japan has a significant lead on peers in active transport and public transit usage rates, this is largely attributed to highly concentrated populations in a few large metropoles like Tokyo, Osaka, and Nagoya. Rural areas and smaller cities have much greater reliance on private cars (Santoso *et al.*, 2012; Abe *et al.*, 2018; Ozaki *et al.*, 2021), and the ageing populations of these regions exacerbates these challenges due to the social and health risks that arise when elderly residents lose the ability to drive. These dynamics can also be seen in some once-wealthy housing estates built during the 1980s and 1990s in peripheral urban areas, which have since become concentrated areas of aged residents left with limited opportunities to access public transport (Ozaki *et al.*, 2021; Sunikka-Blank and Kiyono, 2021). It is thus imperative that Japan avoids these mistakes of the past and concentrates its housing developments in compact cities centred on train and bus routes. Finding solutions to these challenges will enable Japan to offer significant guidance to other nations also facing the imminent challenge of population ageing.

With respect to energy management, Japan's smart cities have exhibited mixed results. HEMS usage studies suggest that providing residents with data on their energy use does little to influence behaviour (Kc *et al.*, 2018; Li *et al.*,

2022). Instead, Japanese people living in homes equipped with HEMS live much as they did before, setting their heat pumps to automatic mode, putting on layers of clothing when cold, and opening the window when warm. Yet, these authors note that homes equipped with HEMS still see lower energy use than homes without these systems. This may be due to smart grid management features beyond user choices, or because HEMS tend to be installed in newer homes built to higher energy efficiency specifications than average in Japan. But there is little indication that HEMS – which cost thousands of USD (hundreds of thousands of JPY) to install – exert an influence on resident energy behaviour that is large enough to rationalise such an investment. The energy efficiency gains of HEMS-equipped homes are also quite modest, with electricity use falling by only 3–7% depending on season (Iwafune *et al.*, 2017), though Yoshida *et al.* (2021) have suggested home fuel cell systems could increase this to about 12%. Given the prominence of these systems in Japan's smart city projects, it is worth asking whether these data-intensive management systems are wise investments compared to simple improvements in building design that reduce energy consumption by increasing thermal performance. Not only does the latter solution offer surer decarbonisation benefits, but it also entails fewer privacy concerns for residents, which are widely reported to have slowed down or derailed ambitions to deploy smart grids or smart homes in other countries (Chen *et al.*, 2020; Del Rio *et al.*, 2021).

5.2 Smart Cities and Established Interests

Are the solutions implemented in Japanese smart cities reflecting the interests of certain firms rather than pursuing the most common sense and low-risk paths to emissions reductions?

All major Japanese smart city projects have entailed partnerships between governments and industry. Private firms have eagerly joined these projects, keen to demonstrate their upcoming technology proposals, test new designs, and secure public funding. Panasonic's efforts in Fujisawa were in part a way to demonstrate their HEMS solutions, including solar panels, batteries, and smart grid technologies (Sakurai and Kokuryo, 2018), while Toyota's Woven City proposals emphasise their decades of investments in fuel cells and autonomous vehicles (Bjarke Ingels Group, 2020). Although corporate players also tend to dominate the conception and implementation of smart cities in other countries (Hollands, 2015; Sadowski and Bendor, 2019), one should recognise that despite their claimed benefits for sustainability, Japan's smart cities are leveraged by the state as a form of industrial policy to promote technological and economic development (Nyberg and Yarime, 2017; DeWit, 2018; Yarime, 2020). Exporting these solutions is a key goal. Japan has several government-run programmes to promote its smart city technologies and businesses overseas, including the Japan Association for Smart Cities in Association of Southeast Asian Nations (JASCA, 2023) and Yokohama's Y-PORT initiative (Y-PORT, 2023).

There is nothing inherently wrong with the Japanese state supporting the interests of domestic industry when pursuing smart city developments. However, to achieve rapid decarbonisation in line with the emission reduction pathways needed to meet Paris Climate Agreement commitments, it is imperative to reduce per-capita energy consumption along with overall material intensity in people's lifestyles (IEA, 2023). As we have shown, under certain conditions, HEMS systems, home PV and battery installation, and autonomous EVs do lead to carbon emissions reductions. We do not dispute the importance of these technologies for accelerating decarbonisation globally. But we should question whether these are the *best available* paths to reducing emissions, especially when they are championed by companies with vested interests in selling their products to sustain their revenue as long as possible. Corporations are not anathema to decarbonisation and can be great partners in development efforts (Sokołowski and Taylor, 2023); however, governments and the public should treat their proposed solutions with caution when simpler alternatives exist.

5.3 Accelerating Decarbonisation Through "Dumber" Choices

Has Japan overlooked "dumb" but more certain solutions to decarbonisation in prioritising smart technologies?

We believe that the close industrial ties of existing Japanese smart city projects have limited the scope of ambition for decarbonisation efforts in previous experiments, due to an over-emphasis on "smart" technologies. Here, we propose "dumb" solutions to some of the challenges of decarbonisation that smart cities are attempting to address in Japan. Each of these solutions is "dumb" because, rather than implementing a cutting-edge technology-based solution, each relies on well-understood social and technically less sophisticated approaches, like the cheapest fuel is the one not used but saved (Sokołowski, 2020), that have been proven over many decades (or centuries).

In the mobility sector, the lowest-emission means of human transport are well known to be walking, cycling, trains, light-rail, and buses. Toyama has attempted to model these solutions already (Ito and Kawazoe, 2022), while Kashiwanoha's compact city and automated bus experiments also embody an attempt to develop public transport rather than personal automobiles (Kashiwa City *et al.*, 2019; Shimono, 2020). Yet other projects like Kitakyushu's support for FCV deployment, Fujisawa's promotion of EV ownership, or Toyota's Woven City still emphasise the role of cars within cities. Work has been done to measure walkability in some Japanese jurisdictions, including Osaka (Kato and Kanki, 2020) and Kashiwanoha (Trencher and Karvonen, 2017). These projects and Japan's historical success with public transport and compact urban spaces also resonate well with the principle of *sufficiency*, emerging from Goldemberg *et al.*'s (1985) work. Swiss urban developments have widely pursued sufficiency by seeking to reduce per-capita energy consumption and avoid the rebound effect by designing apartments with smaller flooring area (Trencher *et al.*, 2018). But there have been no smart city initiatives in Japan that have called for large-scale car-free districts like those proposed in Paris or Brussels (POLIS, 2022). Japan has a long history of having car-free zones on weekends (Ginza, Akihabara, and Shinjuku in Tokyo) or holidays (Kyoto's Gion Matsuri). Yet the country's smart city developments still emphasise the role of personal and shared vehicular mobility in proposed visions of the future city. If decarbonisation is to be a central goal of ongoing smart city developments within Japan, experimenting with car-free development built around the already ubiquitous mixed-use, public transit-based commuter lifestyles of urban Japan would likely provide a surer pathway to reducing GHG emissions.

Such developments would be compatible with efforts to increase green spaces within smart cities. Urban greening is known (Anguluri and Narayanan, 2017) to contribute to reducing urban heat, and thus lowering energy demand for cooling during summer in cities. Several smart cities in Japan have pursued such outcomes, by integrating urban parks and street trees (Okubo *et al.*, 2022). These efforts appear to be continuing, and should. For example, Toyota's Woven City project directly calls for each city block to be built around shared green spaces, including those in both business and residential areas (Bjarke Ingels Group, 2020).

Japan has made significant strides in improving the built environment beyond urban greening. These efforts include building design improvements in existing Japanese smart cities, meeting or exceeding the insulation, natural lighting requirements, and home appliance efficiency standards set at the time each project was built. Yet, while the national government has mandated ZEB and ZEH standards for 2024 and 2030, these standards are not as stringent as some peer jurisdictions. For example, the home thermal insulation standard for ZEH status in Japan is 0.40 kWh/Km^2 (METI, 2015), in the coldest northern climate zone. These standards are less stringent in warmer southern regions like Kyushu and Okinawa than in colder northern areas like Hokkaido and Aomori. But, in January 2023, Scotland committed to mandating that new homes built there meet the international Passivhaus standard (Scottish Government, 2023). Performance requirements for Passivhaus are 0.16 kWh/Km^2 (Passive House Institute, 2015) – 60% lower than the now-binding Japanese standard's strictest region. The construction industry organisation HEAT20 has set higher standards of performance for some custom-designed buildings and homes within Japan, calling for insulation performance of 0.20 kWh/Km^2 in the colder northern regions of Japan and 0.26 kWh/Km^2 in the warmer southern regions (HEAT20, 2020). While less stringent than the Passivhaus standard, HEAT20's standards are 50% lower than the national mandate. Clearly, the Japanese construction industry is capable of significantly outperforming the government's requirements, when customers are willing to pay for these design improvements. This points to an important opportunity for government actors to incentivise efforts by Japanese industry to surpass the newly set thermal performance required by conservative standards.

Japanese smart cities have also overlooked "dumb" solutions to another key problem in the nation's homes: hot water. Yet two such solutions have been well trodden in the country's past. In response to the oil crises of the 1970s, significant investments were made in passive solar water heating installations across Japan through the 1980s and 1990s, with investment only falling off circa the 2008 financial crisis (Farabi-Asl *et al.*, 2019). Before those crises, some geothermal power installations were pursued by local governments, alongside geothermal heat plants (Shortall and Kharrazi, 2017; Hymans and Uchikoshi, 2022). Despite possessing considerable endowments of both solar and geothermal resources for both heat and power, Japan has disproportionately concentrated its investments on solar PV deployment – leading to over 10% of national electricity coming from solar as of 2022 (Institute for Sustainable Energy Policies, 2023). While 35% of home energy use is directed to hot water (Imamura, 2021), in 2015 less than 2% of that heat energy came from solar (Farabi-Asl *et al.*, 2019). Geothermal hot water use is similarly marginal in homes nationwide, yet regionally important. Yet few of Japan's smart cities have engaged with the well-proven technologies of passive solar water heating or geothermal heating. This is despite the studies (Terashima *et al.*, 2020) demonstrating the potential of solar heating technology within Japan and the country's long history with geothermal baths (onsen) as popular parts of daily life and major drivers of tourism (Hymans, 2021; Sokołowski, 2022).

As passive and emission-free means of water heating, solar and geothermal heat could significantly reduce the energy intensity of Japanese homes and lifestyles. For solar hot water systems, these gains could be achieved with something as simple as treated pipes and insulated water tanks – all while maintaining the viability of Japanese cultural bathing practices in a decarbonising world. Geothermal baths, in contrast, are already well established in some regions of Japan for historical reasons, and onsen towns are world-famous tourist attractions. Beppu, Oita, famously allows residents to connect directly to local geothermal spring networks for home onsen use and is exploring further geothermal resource use cases (Endo *et al.*, 2021). There is great potential for both technologies in other parts of Japan. However, harnessing Japan's geothermal resources has proven controversial and is frequently blocked by the influence of residents and lobbyists in onsen towns (Hymans, 2021; Hymans and Uchikoshi, 2022). Yet the emphasis on electrification of water heating present in Japan's smart cities (Okubo *et al.*, 2022) rather than pursuing passive, non-consumptive solutions that had been demonstrated successfully in Japan and elsewhere seems like a significant industry bias. The leading role of Mitsubishi Heavy Industries and Kawasaki in supplying steam generators and related equipment to the global geothermal power industry suggests that Japan also has considerable domestic expertise that could be applied at home to solve some of its culturally relevant energy consumption issues more sustainably.

Finally, the emphasis of Japan's smart communities on smart homes and individual comfort and convenience, rather than environmental performance, has been a subject of critique (Del Rio *et al.*, 2021). Examples include the ongoing encouragement of personal mobility (often advertised as "shared", but sold as mobility as a service by private firms) over collective public transport, home delivery of goods on demand, and risks of the rebound effect. These proposals can each come into conflict with decarbonisation efforts. The convenience of autonomous shared vehicles may not result in emissions reductions. On-demand delivery has been repeatedly found to risk emissions increases compared to traditional batch-order deliveries due to increases in carrier trip frequency (Lin *et al.*, 2018) elsewhere in the world. The rebound effect is particularly well studied, with much evidence suggesting that gains in energy efficiency can drive more energy use through behaviour change over time, eventually outweighing initial energy savings (Muratori and Rizzoni, 2016; de Blas *et al.*, 2018; Stermieri *et al.*, 2023). In the context of smart homes, designing them for user comfort and control, rather than maximising carbon emission reductions, reproduces these risks across Japanese smart cities.

6. Conclusion

Japan has made considerable investments in developing and testing new technologies in its efforts to pursue decarbonisation through smart city projects. However, there are some common features running across many smart cities that pose significant risks to achieving decarbonisation. These chiefly include continued promotion of private automobility rather than expanding Japan's existing culture of modern and ubiquitous public transportation, the roll-out of automated private "smart" homes primarily via new construction rather than retrofitting or upgrading existing building stock, and a focus on demonstrating unproven energy management technologies rather than maximising emissions reductions per se. Each of these tendencies risk hampering decarbonisation outcomes because of embedded carbon in new construction materials, delayed or uncertain emissions reductions benefits, and the availability of alternative and simpler decarbonisation approaches that would likely yield greater returns.

Such risks can be mitigated by reconsidering approaches that rely on cutting-edge "smart" technologies in favour of "dumb" solutions that are already tried and proven. For instance, rather than continuing to design cities around car use, which demands more asphalt and space, smart city planners could emphasise walkability, cycling infrastructure, and public transport. Such a model can be found in Japan's existing pedestrian-only weekend zones that are famous in many core business districts in Tokyo such as Ginza, and Shinjuku. These could be expanded to be year-round car-free zones, a model already successfully implemented in the Yurakucho and Marunouchi Nakadori pedestrianised districts near Tokyo Station. These activities would bring Japan's urban cores in line with other leading cities pursuing a people-centric rather than vehicle-centric future, including Paris and Brussels. Likewise, by emphasising passive designs in homes and buildings, energy consumption could be significantly reduced. Not only would this improve Japan's trade deficit, but it would lessen the need for new grid expansion when pursuing decarbonisation of the electricity sector.

By designing smart cities for minimal consumption from the outset, the emphasis of technologies such as smart grids can be shifted towards a role better served by this technology: collective and regional cooperation across power producers and users. Installing solar panels and batteries in every home may be an appealing vision of a low-carbon future, but it is one that calls for considerable resource investment and recurring replacement and upkeep. It likewise atomises every home, necessitating they each have a bi-directional grid connection to both draw from and feed into the grid, each of which must be capable of withstanding being islanded from the grid during times of overproduction or crisis. The less efficient a nation's buildings are, the more salient these problems become, and the more complex any smart management strategy must be.

This complexity introduces more social, economic, and political risks, especially around the security of the system. For decarbonisation, these risks are most apparent in three realms. The first risk concerns the carbon payback time of new installations (Ren *et al.*, 2023). Unless accompanied by verifiable net-zero building technologies, any new construction or manufacturing necessitates emissions. Further, this verifiability is key to the idea of carbon payback times: any emission benefit from a new product is delayed until it has been used long enough to compensate for the emissions tied to its creation. The second risk arises from the potential to lock-in carbon-intensive systems that do not provide a completely decarbonised solution over the full lifecycle of technology and fuel supply chains. Consider, for instance, fuel cells in FCVs or residential co-generation units (Enefarms) that utilise hydrogen from fossil fuels. Such energy pathways still emit carbon during fuel production, even if coupled with carbon capture (IEA, 2023). The third risk also stems from lock-in, but delays decarbonisation by perpetuating the need for carbon-intensive roads or other infrastructure and by reproducing spread-out cities with detached homes and poor public-transport access (Patella *et al.*, 2019; Sunikka-Blank and Kiyono, 2021; Ward *et al.*, 2021; Ren *et al.*, 2023). Such a residential environment might offer an attractive solution for individual families able to afford the private land and the greater volume of construction materials required for detached homes. But rolling out detached homes – even if smart homes – counters the global shift towards compact cities built around public transport terminals and mixed-use town development. It also runs counter to the historical successes of many cities in Japan that flourished around train stations and concentrated apartment towers, businesses and entertainment facilities in relatively compact land areas. Because

Japan is actively exporting its smart city solutions to other jurisdictions, the potential decarbonisation risks presented by the nation's approach to mobility, housing and energy management should be heeded by city planners in other countries.

"Dumb" solutions can reduce these risks and present new opportunities for export. Developing expertise in designing high-efficiency passive homes, utilising geothermal and solar heating solutions, and minimising total home energy demand would offer Japan massive export opportunities. Today, more than 1.1 billion people (United Nations, 2023) still lack adequate access to energy. Developing new energy supply infrastructure to meet their needs represents a moral and humanitarian necessity, as well as a tremendous business opportunity. There has been much discussion of the potential for the Global South to "leapfrog" the fossil-fuelled development pathway of Japan, the United States, Europe, China, and others. One important contributor to such leapfrogging could be expertise on design strategies that can reduce the amount of energy needed to raise people's standard of living. This would reduce the total energy that must be generated to succeed in "catching up". Japan has accumulated decades of experience demonstrating urban lifestyles and transit modes that are much less energy-intensive than other developed states. Further emphasising these approaches and integrating smart technologies alongside "dumb" ones like renovation, compact development, and passive designs may make Japan's solutions more effective and affordable for overseas partners. With Japan's public and private actors vigorously seeking to export smart city solutions, it is important that Japan offers emerging economies affordable, effective, and low-risk solutions instead of shinier, costlier and resource and land-intensive solutions with uncertain sustainable benefits.

The less risky and "dumber" solutions proposed above point to a need for many of Japan's smart city actors to think outside the box. They also highlight a need to break away from the prevailing tendency to pursue low-carbon development with the limited range of solutions preferred by incumbent private firms. Undoubtedly, reducing vehicle use, retrofitting existing homes instead of building new ones, and doubling down on compact city planning with multifamily units and active or public transportation would create immense carbon and financial savings for consumers. Yet this approach to decarbonisation in smart cities would constitute a far less attractive business proposition to many of the large manufacturers, technology vendors and real-estate developers spearheading many of Japan's smart city developments. Indeed, smaller or new firms would be the most likely to benefit, as they may be more able to pursue new business directions due to less institutional inertia. Established players across Japan's industries may also oppose more decarbonisation-focused "dumb" approaches in smart cities due to the risks of greater competition and market share reduction. Unleashing the decarbonisation potential of smart cities with less risky and less sophisticated solutions would therefore also require considerable leadership from political actors to overcome these conflicting business interests. This leadership would need to ensure that decarbonisation aims are not continuously trumped by industrial policy driven mainly by public and private aspirations to nurture innovation and business opportunities.

Note

1 In Japanese: リノベーション.

References

Abe, T., Seol, J., Kim, M. and Okura, T. (2018) "The Relationship of Car Driving and Bicycle Riding on Physical Activity and Social Participation in Japanese Rural Areas", *Journal of Transport & Health*, 10, pp. 315–321.

Ahvenniemi, H., Huovila, A., Pinto-Seppä, I. and Airaksinen, M. (2017) "What Are the Differences Between Sustainable and Smart Cities?", *Cities*, 60, pp. 234–245.

Akabane, J. (2024) "The Significance of a Smart City for Transformation of an Automaker's Business Model: A Case Study of Toyota's Woven City", in L. Lazzeretti, T. Ozeki, S.R. Sedita and F. Capone (eds) *Clusters in Times of Uncertainty*. Edward Elgar Publishing, pp. 49–66.

Anguluri, R. and Narayanan, P. (2017) "Role of Green Space in Urban Planning: Outlook Towards Smart Cities", *Urban Forestry & Urban Greening*, 25, pp. 58–65.

Autovista24 (2022) "Nissan to Develop Urban Carbon-Neutrality Project in Japanese Smart City". Available at: https://autovista24.autovistagroup.com/news/nissan-develop-carbon-neutrality-project-japan-smart-city/ (Accessed: 30 July 2024).

Barrett, B.F.D., DeWit, A. and Yarime, M. (2021) "Japanese Smart Cities and Communities: Integrating Technological and Institutional Innovation for Society 5.0", in H.M. Kim, S. Sabri and A. Kent (eds) *Smart Cities for Technological and Social Innovation: Case Studies, Current Trends, and Future Steps*. London: Academic Press, pp. 73–94.

Berg, N. (2017) "Raze, Rebuild, Repeat: Why Japan Knocks Down Its Houses After 30 Years", *The Guardian*. Available at: https://www.theguardian.com/cities/2017/nov/16/japan-reusable-housing-revolution (Accessed: 30 July 2024).

Berrill, P., Wilson, E.J.H., Reyna, J.L., Fontanini, A.D. and Hertwich, E.G. (2022) "Decarbonisation Pathways for the Residential Sector in the United States", *Nature Climate Change*, 12, pp. 712–718.

Bjarke Ingels Group (2020) "Toyota Woven City". Available at: https://big.dk/projects/toyota-woven-city-6360 (Accessed: 30 July 2024).

Carlson, J.T. and Trencher, G. (2024) "A Framework for Considering Decarbonisation Risks Emerging From Low-Carbon Hydrogen Supply Chains", *Energy Research & Social Science*, 116, p. 103685.

Chen, C., Xu, X., Adams, J., Brannon, J., Li, F. and Walzem, A. (2020) "When East Meets West: Understanding Residents' Home Energy Management System Adoption Intention and Willingness to Pay in Japan and the United States", *Energy Research & Social Science*, 69, p. 101616.

Choudrie, J., Pheeraphuttranghkoon, S. and Davari, S. (2020) "The Digital Divide and Older Adult Population Adoption, Use and Diffusion of Mobile Phones: A Quantitative Study", *Information System Frontiers*, 22, pp. 673–695.

City of Kitakyushu (2022) "燃料電池自動車(FCV)をはじめとする次世代自動車の有効活用と普及啓発について" [*Regarding the Effective Use and Promotion of Next-Generation Vehicles Including Fuel Cell Vehicles (FCV)*]. Available at: https://www.city.kitakyushu.lg.jp/kankyou/00200146.html (Accessed: 30 July 2024).

City of Yokohama (n.d.) "Large Scale Demonstration with Citizens and Firms: Yokohama Smart City Project (YSCP)". Available at: https://www.city.yokohama.lg.jp/business/

kokusaikoryu/yport/material/pf_jica.files/0039_20181016.pdf (Accessed: 30 July 2024).

Clement, J. and Crutzen, N. (2021) "How Local Policy Priorities Set the Smart City Agenda", *Technological Forecasting and Social Change*, 171, p. 120985.

Cugurullo, F. (2019) "Dissecting the Frankenstein City: An Examination of Smart Urbanism in Hong Kong", in A. Karvonen, F. Cugurullo and F. Caprotti (eds) *Inside Smart Cities: Place, Politics and Urban Innovation*. London: Routledge, pp. 30–44.

de Blas, I., Mediavilla, M., Capellán-Pérez, I. and Duce, C. (2020) "The Limits of Transport Decarbonisation Under the Current Growth Paradigm", *Energy Strategy Reviews*, 32, p. 100543.

de Haan, P., Peters, A. and Scholz, R.W. (2007) "Reducing Energy Consumption in Road Transport Through Hybrid Vehicles: Investigation of Rebound Effects, and Possible Effects of Tax Rebates", *Journal of Cleaner Production, The Automobile Industry & Sustainability*, 15, pp. 1076–1084.

Deguchi, A. (2020) "From Smart City to Society 5.0", in Hitachi-UTokyo Laboratory (ed) *Society 5.0: A People-Centric Super-Smart Society*. Singapore: Springer, pp. 43–65.

Del Rio, D.D.F., Sovacool, B.K. and Griffiths, S. (2021) "Culture, Energy and Climate Sustainability, and Smart Home Technologies: A Mixed Methods Comparison of Four Countries", *Energy and Climate Change*, 2, p. 100035.

DeWit, A. (2014) "Japan's Resilient, Decarbonizing and Democratic Smart Communities", *The Asia Pacific Journal*, 12(50), pp. 2–18.

DeWit, A. (2018) "Japanese Smart Communities as Industrial Policy", in W.W. Clark (ed) *Sustainable Cities and Communities Design Handbook*. Butterworth-Heinemann, pp. 421–452.

Endo, A., Yamada, M., Kenshi, B., Miyashita, Y., Sugimoto, R., Ishii, A., Nishijima, J., Fujii, M., Kato, T., Hamamoto, H., Kimura, M., Kumazawa, T., Masuhara, N. and Honda, H. (2021) "Methodology for Nexus Approach Toward Sustainable Use of Geothermal Hot Spring Resources". *Frontiers in Water*, 3, p. 713000.

Energy Conservation Center Japan (ECCJ) (2022) "The Government Enacted the Revised Building Energy Conservation Act". Available at: https://www.asiaeec-col.eccj.or.jp/policynews-202207/ (Accessed: 30 July 2024).

Farabi-Asl, H., Chapman, A.J., Itaoka, K. and Taghizadeh-Hesary, F. (2019) "Low-Carbon Water and Space Heating Using Solar Energy, Japan's Experience", *Energy Procedia, Innovative Solutions for Energy Transitions*, 158, pp. 947–952.

Fujisawa Sustainable Smart Town (2023) "Introducing Fujisawa Sustainable Smart Town". Available at: https://fujisawasst.com/EN/wp_en/wp-content/themes/fujisawa_sst/pdf/FSST-ConceptBook.pdf (Accessed: 30 July 2024).

Gazoo (2020) *100年続く街を目指す神奈川県・藤沢のスマートシティ。新しいことを発信して「世界中に広げたい」—モビリティを取り巻くサービスの展望④* [*A Smart City in Fujisawa, Kanagawa Prefecture That Aims to be a City That Will Last for 100 Years. We Want to Communicate New Things and Spread Them Around the World – Prospects for Services Surrounding Mobility ④*]. Available at: https://gazoo.com/mobility/maas/20/03/28/ (Accessed: 30 July 2024).

Goldemberg, J., Johansson, T., Williams, R.H. and Reddy, A.K.N. (1985) "Basic Needs and Much More With One Kilowatt Per Capita", *Ambio*, 14, pp. 190–200.

Granier, B. and Kudo, H. (2016) "How Are Citizens Involved in Smart Cities? Analysing Citizen Participation in Japanese 'Smart Communities'", *Information Polity*, 21, pp. 61–76.

Gurtu, A. and Johny, J. (2021) "Supply Chain Risk Management: Literature Review", *Risks*, 9(1), p. 16.

Harvey, J., Guo, W. and Edwards, S. (2019) "Increasing Mobility for Older Travellers Through Engagement With Technology", *Transportation Research Part F: Traffic Psychology and Behaviour*, 60, pp. 172–184.

HEAT20 (2020) HEAT20 住宅シナリオ [*HEAT20 Housing Scenario*]. Available at: http://www.heat20.jp/grade/index.html (Accessed: 30 July 2024).
Hertz, J. (2023) "How One Japanese Smart City is Bringing Wireless EV Charging to the Road", *EE Power*. Available at: https://eepower.com/market-insights/how-one-japanese-smart-city-is-bringing-wireless-ev-charging-to-the-road/# (Accessed: 30 July 2024).
Hollands, R.G. (2015) "Critical Interventions into the Corporate Smart City. Cambridge Journal of Regions", *Cambridge Journal of Regions, Economy and Society*, 8(1), pp. 61–77.
Hornyak, T. (2022) "Why Japan is Building Smart Cities From Scratch", *Nature*, 608, S32–S33.
Hymans, J.E.C. (2021) "Losing Steam: Why Does Japan Produce So Little Geothermal Power?", *Social Science Japan Journal*, 24(1), pp. 45–65.
Hymans, J.E.C. and Uchikoshi, F. (2022) "To Drill or Not to Drill: Determinants of Geothermal Energy Project Siting in Japan", *Environmental Politics*, 31, pp. 407–428.
Imamura, T. (2021) "Japan's Roadmap to Carbon Neutrality in the Building and Housing Sectors", *The 7th IEA-Tsinghua Joint Workshop*, 30 November 2021. Available at: https://iea.blob.core.windows.net/assets/993f67a2-054e-41ba-9adb-f782358a4e73/3.JapansRoadmaptoCarbonNeutralityintheBuildingandHousing-Sectors.pdf (Accessed: 30 July 2024).
Institute for Sustainable Energy Policies (2023) "2022 Share of Electricity From Renewable Energy Sources in Japan (Preliminary)". Available at: https://www.isep.or.jp/en/1436/ (Accessed: 30 July 2024).
International Energy Agency (2021) "Global Hydrogen Review 2021". Available at: https://www.iea.org/reports/global-hydrogen-review-2021 (Accessed: 30 July 2024).
International Energy Agency (2023) "Net Zero Roadmap: A Global Pathway to Keep the 1.5°C Goal in Reach – 2023 Update". Available at: https://www.iea.org/reports/net-zero-roadmap-a-global-pathway-to-keep-the-15-0c-goal-in-reach (Accessed: 30 July 2024).
Ito, H. and Kawazoe, N. (2022) "Promoting Urban Light Rail Transit in a Compact City Context: The Case of Toyama City, Japan", *Regional Studies, Regional Science*, 9, pp. 776–793.
Itokawa, S. (2020) "Comparison of Cities' Transportation Modal Shares and Post-Coronavirus Projections", *SC-ABeam Automotive Consulting*. Available at: https://www.sc-abeam.com/and_mobility/en/article/20201203-01/ (Accessed: 30 July 2024).
Iwafune, Y., Mori, Y., Kawai, T. and Yagita, Y. (2017) "Energy-Saving Effect of Automatic Home Energy Report Utilizing Home Energy Management System Data in Japan", *Energy*, 125, pp. 382–392.
Japan Association for Smart Cities in ASEAN (JASCA) (2023) "About JASCA". Available at: https://www.jasca2021.jp/about/ (Accessed: 30 July 2024).
Kashiwa City, Mitsui Fudosan Co. and Urban Design Center Kashiwa-no-ha (2019) "Selected by the Ministry of Land, Infrastructure, Transport and Tourism as a Model Project for Smart City Towards Realizing 'Society 5.0' Kashiwa-no-ha Smart City Will Evolve into a Smart, Compact City Driven by Data". Available at: https://www.mitsuifudosan.co.jp/english/corporate/news/2019/0605_02/download/20190605.pdf.
Kato, H. and Kanki, K. (2020) "Development of Walkability Indicator for Smart Shrinking", *International Review for Spatial Planning and Sustainable Development*, 8(1), pp. 39–58.
Kc, R., Rijal, H., Shukuya, M. and Yoshida, K. (2018) "An In-Situ Study on Occupants' Behaviors for Adaptive Thermal Comfort in a Japanese HEMS Condominium", *Journal of Building Engineering*, 19, pp. 402–411.
Kitchin, R., Coletta, C. and Heaphy, L. (2019) "Actually Existing Smart Dublin: Exploring Smart City Development in History and Context", in A. Karvonen, F. Cugurullo and

Caprotti, F. (eds) *Inside Smart Cities: Place, Politics and Urban Innovation*. London: Routledge, pp. 85–101.

Kusanagi, S. (2023) "Smart Energy Networks, Decarbonization and Public Utility Infrastructure Resilience: The Case of Japan", in M.M. Sokołowski and A. Visvizi (eds) *Routledge Handbook of Energy Communities and Smart Cities*. London: Routledge, pp. 249–261.

Li, Y., Gao, W., Zhang, X., Ruan, Y., Ushifusa, Y. and Hiroatsu, F. (2020) "Techno-Economic Performance Analysis of Zero Energy House Applications with Home Energy Management System in Japan", *Energy and Buildings*, 214, p. 109862.

Li, Y., Zhang, X., Gao, W. and Qiao, J. (2022) "Lessons Learnt From the Residential Zero Carbon District Demonstration Project, Governance Practice, Customer Response, and Zero-Energy House Operation in Japan", *Frontiers in Energy Research*, 10, p. 915088.

Lin, J., Zhou, W. and Du, L. (2018) "Is On-Demand Same Day Package Delivery Service Green?" *Transportation Research Part D: Transport and Environment*, 61(Part A), pp. 118–139.

Luque-Ayala, A. and Marvin, S. (2015) "Developing a Critical Understanding of Smart Urbanism?", *Urban Studies*, 52, pp. 2105–2116.

Martin, C.J., Evans, J. and Karvonen, A. (2018) "Smart and Sustainable? Five Tensions in the Visions and Practices of the Smart-Sustainable City in Europe and North America", *Technological Forecasting and Social Change*, 133, pp. 269–278. https://doi.org/10.1016/j.techfore.2018.01.005.

METI (2015) "ZEH Roadmap Review Committee: Summary", *National Diet Library*. Available at: https://warp.da.ndl.go.jp/info:ndljp/pid/9766998/www.meti.go.jp/press/2015/12/20151217003/20151217003-1.pdf (Accessed: 30 July 2024).

METI (2023) "Cabinet Decision on the Basic Policy on the Rationalizing Use of Energy and Shifting to Non-Fossil Energy". Available at: https://www.meti.go.jp/english/press/2023/0317_003.html (Accessed: 30 July 2024).

Moriarty, P. and Wang, S.J. (2017) "Could Automated Vehicles Reduce Transport Energy?", *Energy Procedia, Proceedings of the 9th International Conference on Applied Energy*, 142, pp. 2109–2113.

Muratori, M. and Rizzoni, G. (2016) "Residential Demand Response: Dynamic Energy Management and Time-Varying Electricity Pricing", *IEEE Transactions on Power Systems*, 31, pp. 1108–1117.

Nagata, K. (2023) "Japan's Scrap-and-Rebuild Culture Faces an Environmental Reckoning", *The Japan Times*. Available at: https://www.japantimes.co.jp/environment/2023/08/27/sustainability/japan-scrap-build-sustainability/ (Accessed: 30 July 2024).

Nissan (2021) "Reinventing Local Transport and Energy Supply: All in One Project". Available at: https://www.nissan-global.com/EN/STORIES/RELEASES/namie/ (Accessed: 30 July 2024).

Nyberg, A.R. and Yarime, M. (2017) "Assembling a Field into Place: Smart-City Development in Japan", in M.-D.L. Seidel and H.R. Greve (eds) *Emergence*. Leeds: Emerald, pp. 253–279.

Okubo, H., Shimoda, Y., Kitagawa, Y., Gondokusuma, M.I.C., Sawamura, A. and Deto, K. (2022) "Smart Communities in Japan: Requirements and Simulation for Determining Index Values", *Journal of Urban Management*, 11, pp. 500–518.

Ozaki, R., Aoyagi, M. and Steward, F. (2021) "Community Sharing: Sustainable Mobility in a Post-Carbon, Depopulating Society", *Environmental Sociology*, 8, pp. 73–87.

Passive House Institute (2015) "Passive House Requirements". Available at: https://passivehouse.com/02_informations/02_passive-house-requirements/02_passive-house-requirements.htm (Accessed: 30 July 2024).

Patella, S.M., Scrucca, F., Asdrubali, F. and Carrese, S. (2019) "Carbon Footprint of Autonomous Vehicles at the Urban Mobility System Level: A Traffic Simulation-Based

Approach", *Transportation Research Part D: Transport and Environment*, 74, pp. 189–200.
POLIS (2022) "Brussels and Paris to Establish Largely Car-Free City Centres", *POLIS: Cities and Regions for Transport Innovation*. Available at: https://www.polisnetwork.eu/news/brussels-and-paris-to-establish-largely-car-free-city-centres/ (Accessed: 30 July 2024).
Ren, Y., Sun, X., Wolfram, P., Zhao, S., Tang, X., Kang, Y., Zhao, D. and Zheng, X. (2023) "Hidden Delays of Climate Mitigation Benefits in the Race for Electric Vehicle Deployment", *Nature Communications*, 14, p. 3164.
Renoveru (2019) "Using Renovation to Create a Recycling-Oriented Society, One With Smart, Enjoyable Home Lives, Characterized by Well-Being". Available at: https://renoveru.co.jp/en/about/sustainability/ (Accessed: 30 July 2024).
Sadowski, J. and Bendor, R. (2019) "Selling Smartness: Corporate Narratives and the Smart City as a Sociotechnical Imaginary", *Science, Technology, & Human Values*, 44, pp. 540–563.
Sakuma, N., Trencher, G., Yarime, M. and Onuki, M. (2021) "A Comparison of Smart City Research and Practice in Sweden and Japan: Trends and Opportunities Identified From a Literature Review and Co-Occurrence Network Analysis", *Sustainability Science*, 16, pp. 1777–1796.
Sakurai, M. and Kokuryo, J. (2018) "Fujisawa Sustainable Smart Town: Panasonic's Challenge in Building a Sustainable Society", *Communications of the Association for Information Systems*, 42(1), pp. 508–525.
Santoso, D.S., Yajima, M., Sakamoto, K. and Kubota, H. (2012) "Opportunities and Strategies for Increasing Bus Ridership in Rural Japan: A Case Study of Hidaka City", *Transport Policy*, 24, pp. 320–329.
Schwartz, Y., Raslan, R. and Mumovic, D. (2022) "Refurbish or Replace? The Life Cycle Carbon Footprint and Life Cycle Cost of Refurbished and New Residential Archetype Buildings in London", *Energy*, 248, p. 123585.
Scottish Government (2023) "Energy Standards Review – Scottish Passivhaus Equivalent: Working Group". Available at: https://www.gov.scot/groups/energy-standards-review-scottish-passivhaus-equivalent-working-group/ (Accessed: 30 July 2024).
Shimono, K. (2020) "Long Term Pilot Deployment for Automated Driving Bus Operation at Kashiwa-no-ha Area". Available at: https://en.sip-adus.go.jp/evt/workshop2020/file/ra/04RA_03E_Shimono.pdf (Accessed: 30 July 2024).
Shortall, R. and Kharrazi, A. (2017) "Cultural Factors of Sustainable Energy Development: A Case Study of Geothermal Energy in Iceland and Japan", *Renewable and Sustainable Energy Reviews*, 79, pp. 101–109.
Sokołowski, M.M. (2020) *European Law on Combined Heat and Power*. London: Routledge.
Sokołowski, M.M. (2021) "Models of Energy Communities in Japan (Enekomi): Regulatory Solutions From the European Union (Rescoms and Citencoms)", *European Energy and Environmental Law Review*, 30(4), pp. 149–159.
Sokołowski, M.M. (2022) *Energy Transition of the Electricity Sectors in the European Union and Japan: Regulatory Models and Legislative Solutions*. Cham: Palgrave Macmillan.
Sokołowski, M.M. and Taylor, M. (2023) "Just Energy Business Needed! How to Achieve a Just Energy Transition by Engaging Energy Companies in Reaching Climate Neutrality: (Re)Conceptualising Energy Law for Energy Corporations", *Journal of Energy & Natural Resources Law*, 41(2), pp. 157–174.
Stermieri, L., Kober, T., Schmidt, T.J., McKenna, R. and Panos, E. (2023) "Quantifying the Implications of Behavioral Changes Induced by Digitalization on Energy Transition: A Systematic Review of Methodological Approaches", *Energy Research & Social Science*, 97, p. 102961.

Sunikka-Blank, M. and Kiyono, Y. (2021) "Why Do You Need More Towers? Four Approaches to Sustainable Urban Regeneration in Japan", *ARQ: Architectural Research Quarterly*, 25(4), pp. 372–383.

Tada, N., Marui, M. and Mizutani, A. (2014) "Promotion of Smart Community in Aizuwakamatsu City Area", *FUJITSU Science Technical Journal*, 50(2), pp. 11–18. Available at: https://www.fujitsu.com/global/documents/about/resources/publications/fstj/archives/vol50-2/paper02.pdf (Accessed: 30 July 2024).

Terashima, K., Sato, H. and Ikaga, T. (2020) "Development of an Environmentally Friendly PV/T Solar Panel", *Solar Energy*, 199, pp. 510–520.

Trencher, G. (2019) "Towards the Smart City 2.0: Empirical Evidence of Using Smartness as a Tool for Tackling Social Challenges", *Technological Forecasting and Social Change*, 142, pp. 117–128.

Trencher, G., Geissler, A. and Yamanaka, Y. (2018) "15-Years and Still Living: The Basel Pilot Region Laboratory and Switzerland's Pursuit of a 2000-Watt Society", in S. Marvin et al. (eds) *Urban Living Labs: Experimentation and Socio-Technical Transitions*. London: Routledge, pp. 167–188.

Trencher, G. and Karvonen, A. (2017) "Stretching 'Smart': Advancing Health and Well-Being Through the Smart City Agenda", *Local Environment*, 24, pp. 610–627.

United Nations (2023) "Energy Access and Affordability: Powering Ahead to 2030". Available at: https://unsdg.un.org/latest/stories/energy-access-and-affordability-powering-ahead-2030 (Accessed: 30 July 2024).

Wadud, Z., MacKenzie, D. and Leiby, P. (2016) "Help or Hindrance? The Travel, Energy and Carbon Impacts of Highly Automated Vehicles", *Transportation Research Part A: Policy and Practice*, 86, pp. 1–18.

Ward, J.W., Michalek, J.J. and Samaras, C. (2021) "Air Pollution, Greenhouse Gas, and Traffic Externality Benefits and Costs of Shifting Private Vehicle Travel to Ridesourcing Services", *Environmental Science & Technology*, 55(19), pp. 13174–13185.

Wuyts, W., Miatto, A., Sedlitzky, R. and Tanikawa, H. (2019) "Extending or Ending the Life of Residential Buildings in Japan: A Social Circular Economy Approach to the Problem of Short-Lived Constructions", *Journal of Cleaner Production*, 231, pp. 660–670.

Yagita, Y. and Iwafune, Y. (2021) "Residential Energy Use and Energy-Saving of Older Adults: A Case From Japan, the Fastest-Aging Country", *Energy Research & Social Science*, 75, p. 102022.

Yarime, M. (2020) "Facilitating Innovation for Smart Cities: The Role of Public Policies in the Case of Japan", in Y.-M. Joo and T.-B. Tan (eds) *Smart Cities in Asia: Governing Development in the Era of Hyper-Connectivity*. Cheltenham: Edward Elgar Publishing, pp. 93–106.

YKKAP (n.d.) "Taking on the Labour Shortage Problem". Available at: https://www.ykkap.co.jp/company/en/sustainability/story/labor-shortage.html (Accessed: 30 July 2024).

Yoshida, K., Rijal, H.B., Bohgaki, K., Mikami, A. and Abe, H. (2021) "Energy-Saving and CO2-Emissions-Reduction Potential of a Fuel Cell Cogeneration System for Condominiums Based on a Field Survey", *Energies*, 14, p. 6611.

Y-PORT (Yokohama Partnership of Resources and Technologies) (2023) "Innovative Urban Solutions by Yokohama". Available at: https://yport.city.yokohama.lg.jp/en/innovative-urban-solutions-developed-by-yokohama (Accessed: 30 July 2024).

Zhou, Y., Sato, H. and Yamamoto, T. (2023) "Resourcing Idle Privately Owned Autonomous Electric Vehicles: System Design and Simulation for Ride-Sharing in Residential Areas of the Aging in Japan", *Transport Policy*, 143, pp. 46–57.

5 Deepening the City-Region Gap in 21st Century Japan

Smart Cities as a Tool to Achieve Administrative Neoliberalisation

Kie Sanada and Marco Zappa

1. Introduction

In 21st-century Japan, a renewed emphasis is put on the need to develop the digital and infrastructure sectors as the country grapples with a demographic crisis and regional competition. In 2022, the Liberal Democratic Party (LDP)-led Cabinet under Prime Minister Fumio Kishida, for instance, launched a Digital Garden City Nation (DGCN) plan aiming to spur "bottom-up growth" through massive public and private investments in the modernisation of both urban and rural areas through the creation of a plethora of smart cities (Kishida, 2022). Tokyo officially defines a smart city as "a sustainable city or region"[1] where information and communication technologies (ICT) implementation allows the upgrade of local management, the resolution of local issues, and the continuous generation of new values, and hence, "a place for the prompt materialisation of the Society 5.0" (Cabinet Secretariat, 2023, p. 10).[2] This was to set forth on and reinforce his predecessors' path. Particularly, during the late period of his long term as Japan's Prime Minister (2012–2020), Shinzō Abe made great efforts to promote the idea of "Society 5.0", or a "society based on future technologies",[3] where ICT and artificial intelligence (AI)-based solutions (drones, robots, etc.) are implemented in various sectors such as agriculture, manufacturing, service, and logistics, to foster rural revitalisation and contribute to the locales' own self-reliance in region-wide "Circulating and Ecological" economies (Sokołowski, 2021b; Tokunaga and Shimobe, 2021). The post-Abe Suga administration added decarbonisation as a discursive layer. DGCN promises that its successful implementation would eventually lead Japan to meet its ambitious decarbonisation targets – 46% reduction of greenhouse gas (GHG) emissions from the 2013 levels and complete carbon neutrality by 2050 – and improve the nation's quality of life (QOL).

Hence, there is an insistence on smart cities, particularly in the energy and mobility sectors, as a key component of the Society 5.0 strategy. If, on the one hand, the central government sets the agenda at the national level, then on the other, local governments[4] are responsible for drafting proposals, gathering finances, and implementing specific projects addressing major issues at the

DOI: 10.4324/9781003471448-5

local level (e.g. energy production, supply, efficiency, depopulation). In this context, we intend to address the following research questions: why has an increasing number of local administrations throughout Japan recurred to the smart city concept, broadly defined, to promote environmental sustainability, energy efficiency, and urban renovation, and to attract investments and visitors thus enhancing place branding?

Taking a multi-level perspective, in this chapter we aim at clarifying the role of multiple actors involved in a complex "sociotechnical network" underlying the current emphasis on smart cities as a policy idea in Japan. Our main hypothesis is that the structural reforms of the 1990s and 2000s have succeeded in encouraging communities to support themselves, reducing their dependence on the central government's financial transfers. To support our argument, we proceeded with a careful mapping of the smart city initiatives and frameworks in Japan in the last decade by adopting the Urban Employment Area (UEA) framework developed by Kanemoto and Tokuoka (2002). Our results show that several localities have increased their influence and power vis-à-vis Tokyo and prefectural governments. In fact, we argue that in this wave of neoliberalisation of local governance, only where a certain degree of resources already existed (mainly in urban areas), have local administrations maintained and expanded their power vis-à-vis the central government. By contrast, particularly in non-urban areas, *jichitai* are struggling to stay afloat, even though they resort to smart city policies. In the following sections, we will proceed first to illustrate our theoretical and analytical framework.

2. Theoretical and Analytical Framework

The dominant narrative on smart cities as a place where citizens can achieve happiness and well-being in so far as they have access to a wide array of ICT-based services or automation of urban processes has so far proven far-fetched and at times misleading (Greenfield, 2017; Gonella, 2019). Adopting a sociological view on technology, it is clear that "technology has no power of itself (. . .) but "[o]nly in association with human agency, social structures and organisations, fulfil its functions" (Geels, 2002). In other words, technologies become embedded in routines, patterns of behaviour, regulations, and so on, becoming themselves actors (though non-human) in a wide network of interactions (Latour, 1990).

To borrow again from Geels (2002), smart cities, interpreted as ICT-based residential communities, are not just an assemblage of innovation applied to urban life. Rather, they are at the centre of specific "sociotechnical configuration(s)", institutionalised in a locally specific manner. In fact, the strive for city liveability and resilience calls into question several intertwined issues that need to be addressed systematically and from a multi-level perspective. In their seminal edited book, Sokołowski and Visvizi (2023) offer an in-depth discussion of the challenges and opportunities surrounding energy communities, i.e. cooperative and citizen-centred forms of energy production and distribution expedited by advanced technology, are increasingly playing in the definition of smart cities around the globe. In fact, the development of energy communities is a key element of a "socio-environmental transition" towards more democratic and citizen-centred urban models, or towards an "extension of the 'autonomy' of urban agents in a system of organised complexity" (De Franco *et al.*, 2023, p. 20). Nonetheless, particularly in the Japanese context, smart city planning, in spite of its energy focus, has emerged since the mid-2000s as an item in the administrative and regulatory strategies of both central and local governments that have ended up suffocating more radical self-governance and autonomy drives while supporting the neoliberal turn of local governance since the mid-1990s and early 2000s. Following Morozov and Bria (2018), we define neoliberalisation as a tendency towards "decentralised governance" enabled through specific mechanisms and technologies, such as the adoption of rankings measuring city competitiveness or indebtedness based on which investors decide whether to invest in one place or not; the integration of audit mechanisms into local governance and the increasing importance of data collecting, analysing and processing; the emergence of a speculative logic underpinning infrastructure construction and maintenance (Morozov and Bria, 2018, pp. 8–12).

In this light, smart cities are construed as places where the use of ICT makes urban activities more efficient and sustainable economically, socially, and environmentally; this has gradually emerged as a leading policy idea embodying a technology-based one-fit-all solution to urban problems in advanced economies (Gonella, 2019). In the sociotechnical configuration framework, when a smart city is materialised, technical upgrades of city management are accompanied by social innovations. Here, social innovation is understood as the alteration of practices, which occurs due to newly generated values via access to new information, technologies, ideas, and services (Baron *et al.*, 2018). This points to institutional changes centred on the redefinition of historically constituted, locally specific structures of moral responsibilities, through which agents regulate and enact practical judgements.

Accordingly, the materialisation of smart innovation is understood in three stages. First, it can occur in terms of the introduction of technological solutions to increase the efficiency and productivity of public management. It can be adopted within a limited timeframe for the purpose of technical experiments and/or for a longer term as a part of the governance process. This stage can involve the digital upgrade of existing operations. In fact, most urban infrastructural projects do *smart*ify the existing operation but do not necessarily accompany changes in terms of socio-political configuration as well as the citizens' livelihood.

Second, through access to new technology, a redefinition of the roles held by public authorities, private businesses, and civil society in local governance may occur. In the process of smart city building, public bodies must rely more or less on private partnerships to compensate for the lack of expertise in implementing technological- and digital-infrastructure and data-oriented solutions (Voorwinden, 2021). The new configuration of stakeholders may trigger a redefinition of moral boundaries of democratic governing responsibilities, which is reflected in and represented by the innovativeness of local smart city planning.

Third, the new digitally informed advanced technologies can reshape the existing workings of public management (e.g. waste, mobility, administrative management) as a whole. This leads to the establishment of a new way of regulating and monitoring practices. Indeed, this stage exerts concrete impacts on citizens by altering daily operations in a smarter way. Existing studies on Japanese smart cities take into consideration smart city prototypes widely publicised by Japanese authorities, such as Aizuwakamatsu and Kitakyūshū(DeWit, 2014; Trencher, 2019; Sakuma *et al.*, 2021). In this sense, it is possible to identify technocratic tendencies when it comes to smart city planning and development (Zappa, 2020). From the standpoint of the three stages of smart innovation materialisation elucidated above, these cases are however exemplary of the first stage only (see Table 5.1).

Through our study, we demonstrate that, by contrast, less publicised cases, particularly from remote areas, some of which have been visited by one of the authors of this chapter, vividly show the accomplishment of the second stage of smart innovation in terms of structural administrative change towards the materialisation of "Society 5.0" proposed by the Government of Japan. In the following sections, we discuss how structural reforms since the early 2000s have changed (or at least attempted to do so) local governance and how the smart city policy idea comes into play in this context promoting the neoliberalisation of local governance.

Table 5.1 The stages of smart innovation materialisation

Stage	*Policy*	*Aim*	*Duration*
First	State-led technological infrastructure upgrade	Operation optimisation	Short term; experimental
Second	Subcontracting use of infrastructure, public-private partnerships (PPPs)	Compensation for lack of expertise	Mid- to long term; may lead to redefinition of roles within governance
Third	Extension of technological application to various public services	Comprehensive public sector reform	Long term; modifications to citizens' daily lives

3. A Managerial Approach to Local Governance: Decentralisation and the Role of *Jichitai*

In this section, we focus on the Government of Japan's and local governments' initiatives between the 2010s and 2020s. Particularly, Tokyo's reform initiatives towards administrative decentralisation and the role of local governments will be discussed.

Following Hajer's analytical framework (Hajer, 2005), it is possible to historically situate the emergence of a "discourse coalition", that is a group of actors sharing a common definition for what was, and still is, a rather broad and opaque phenomenon. Several factors, both contingent and structural, might be pointed out as contributing to the historical emergence of said discourse coalition in Japan's political realm.

First, the smart city concept drew nationwide attention in the aftermath of the 11 March 2011 triple disaster of Northeastern Japan (3/11) which forced the Government of Japan to adopt specific policies aimed at enhancing the resilience of both urban and rural communities particularly in terms of energy generation and distribution (Sokołowski, 2022, p. 197). Damages and disruptions caused by the 3/11, in fact, highlighted the vulnerability of Japan's regionalised energy generation and supply infrastructure and pushed policymakers to find new solutions to decade-long problems (Zappa, 2020, pp. 198–199).

On top of disaster resilience, energy conservation and self-sufficiency have been a long-term target of the Government of Japan since the oil crises in the mid- and late 1970s. In light of these external shocks, Tokyo launched two national programmes (the Sunshine Program and the Moonlight Program) aimed at fostering public–private cooperation in the energy sector to find feasible alternatives to oil, stressing the importance of the development of new technologies that allow better energy efficiency and conservation (Kimura, 2009). Japanese authorities also drafted specific legislation, namely the 1979 Act on Rationalisation of Energy Use[5] which required 90% of business operators at that time to reduce their use of fossil-derived fuels (mainly oil, gas and coal) by 1% and, more significantly, introduced minimum energy performance standards for vehicles, appliances, and buildings (Sokołowski, 2022, pp. 189–190). Building on this foundation, amidst growing concern surrounding the effects on Japan's ecosystems and populace of climate change and global warming, post-bubble Heisei Era cabinets in the 1990s and early 2000s took steps to further promote cuts to GHG emissions and environmental energy efficiency standards. In 2005, Tokyo adopted a Kyoto Protocol Target Achievement Plan to achieve a 6% reduction of GHG emissions as agreed during the 1997 COP3 in Kyoto (Deguchi, 2020). It is noteworthy that Aichi Expo in the same year centred around the relation between humans and nature, and served both as a display of and catalyst to Japanese innovations in domains such as energy and mobility through the use of renewable energy sources, such as solar, biogas, and hydrogen (Zappa, 2020, p. 201).

Lastly, the realisation of smart city initiatives is integral to the process of *administrative decentralisation* begun in the early 2000s with the structural reforms promoted by the then Prime Minister Junichirō Koizumi. Centred on the so-called trinity,[6] Koizumi's plan envisaged a reduction of and, where possible, a complete halt to central government subsidies to local governments, the promotion of tax income transfer, and a sweeping revision of the taxes allocated to local governments.

Despite the reforms being generally considered to have fallen short of their targets (Niimura, 2018, p. 101), several administrative procedures, such as central government approval in the relevant area of city planning, have been curtailed resulting in the transfer of more responsibilities to local assemblies. At the same time, these initiatives have not properly made up for the growing demand of financial resources caused by a partial devolution of powers, spurring, if anything, a drive towards inter-prefectural competition for resource attraction.

From a neoliberal perspective, the obvious long-term aim of these reforms was to disenfranchise local administrations from their dependence on financial transfers from the central government through a broader engagement with the private sector and external funding entities. Concomitantly, through audit and evaluation mechanisms, it created a classification and reward system for virtuous local governments adhering to the Government of Japan's guidelines in any relevant policy sector.

The Constitution of Japan in Articles 92–94 recognise the principle of local autonomy, the right for citizens to elect their prefecture governor democratically, and give the local governments the power to manage properties, affairs, and administration and to enact their own regulations within the law. With the enactment of the structural reforms, in addition to the above, local governments were entrusted with the capacity to write and submit funding proposals to the central government within a certain funding framework. To guarantee the success of the application, project proposals need to be carved along standards that are defined by the central government, and particularly by the Cabinet Secretariat, which oversees and monitors the government's initiatives in the area of regional revitalisation.

This model of decentralised governance has become embedded in the subsequent subsidy schemes devised by the Government of Japan to push ahead its local revitalisation plans before and after 3/11. In 2008, for instance, the Fukuda's administration launched its Environmental Model City Plan encouraging local administrations to take concrete actions locally to cut carbon emissions in exchange for the government's designation as "environmental model cities"[7] and material and financial support to the technological transition for GHG emissions reduction (Sugiyama and Takeuchi, 2008, pp. 428–429; Zappa, 2020, p. 201).

Since 2014, with the entering into force of the City, People, Job Creation Law (Government of Japan, 2014),[8] the Government of Japan has stepped up its efforts to promote local economic development and revitalisation drafting successive medium-term comprehensive strategies[9] that it expects to be reflected in single local authorities' growth plans.[10] At the core of these instructions was the urge to spur "local Abenomics",[11] to promote innovation, place branding, and the productivity of local services through a mix of financial, educational, and informational support, with the ambitious aim to "create a new inflow of people in the regions" (Jichitai Tsūshin Online, 2020).[12]

Once submitted, plans and strategies of *jichitai* are assessed and monitored by ad hoc expert committees (Morita *et al.*, 2020). To facilitate the evaluation and monitoring procedure, standard managerial tools and concepts such as the plan-do-check-act (PDCA) cycle are adopted to stress the individual project's strengths and weaknesses and are used to promote continuous improvement[13] of the performance at the local level. This, in turn, is the subject of a series of key performance indicators representing the specificities of individual projects or local contingencies (Jichitai Tsūshin Online, 2020).

Considering the above, the Cabinet Secretariat provides guidelines and manuals through its website, thus providing support and counselling to local administrators throughout the submission process. For instance, local governments are encouraged to create their own networks across sectors of society and beyond prefectural and regional boundaries. In one of such guides compiled by the Cabinet Secretariat City, People, and Job Creation Bureau, local administrators are urged to engage with external experts and young people in the deliberative phase and to benchmark with other local administrations (Cabinet Secretariat *et al.*, 2021).

In this context, we cannot avoid submitting the hypothesis that urban areas, having access to a broader pool of resources (particularly in terms of financial reserves, possibility to attract private capitals and highly skilled labour) have benefitted the most, attracting new flows of people and capital from both Japan and abroad (Nakazawa, 2014). In the attempt to reduce the urban-rural divide, the Government of Japan has increasingly resorted to promoting ICT utilisation for the improvement of the QOL launching subsidy schemes and public awareness initiatives revolving around the concept of the smart city. Against the backdrop of a sluggish pace of installation of optical fibre and 5G in remote areas, the cabinets of both Suga and Kishida have stepped up their efforts in promoting digitalisation and infrastructure modernisation.

Excluding a few notable exceptions,[14] Japanese smart cities have emerged as *brownfield* projects; that is, they are village or neighbourhood community-wide model areas within larger regions or urban areas usually functioning as a showcase for technological advance in a wide array of sectors (housing, mobility, energy, agriculture, etc.) vis-à-vis domestic or international visitors. In other words, Japanese smart cities are mostly experimental in nature, though comprehensive platforms built to respond to a wide array of issues, ranging from energy to demographic decline, have emerged (Barrett *et al.*, 2021).

Certainly, until 3/11 environmental and energy issues have served as the dominant framework within which policy could be articulated. However, the importance of other issues, such as disaster preparedness, traffic reduction, and waste management, shall not be overlooked. Under the second Abe administration (2012–2020), for instance, such efforts have been expanded under the banner of the "Society 5.0". Since the mid-2010s, "microgrids", i.e. local self-sufficient electric power networks supported by small-scale power generators (usually photovoltaic), storage units, and ICT-based energy management systems (smart metres, home energy management systems [HEMS], etc.) allowing for energy efficiency and CO_2 reduction have emerged as the dominant form of smart city in Japan. Their success can be described as follows: on the one hand, they are central to the promotion of local innovation and disaster preparedness,[15] while on the other hand, they could fit into the Government of Japan's regional revitalisation targets.

Large companies such as Panasonic and Toyota have been directly engaged in the establishment of the so-called smart communities based on the aforementioned microgrids. The Fujisawa Sustainable and Smart Town (Kanagawa Prefecture) is a notable example. With a cluster of two- to three-story "smart" housing units with photovoltaic panels installed on the roof, surplus energy storage facilities (usually solid oxide fuel cell batteries), electric vehicles' charging facilities, HEMS, and energy data management system connected to the Internet, built on a former Panasonic plant, it is almost entirely powered by renewable energy sources (70% of the total energy demand) and has reduced CO_2 emissions, ensuring energy reserves in case of power outages.

More recently, community-led smart communities have emerged (Sokołowski and Visvizi, 2023). Flagged as a successful revitalisation project by the Ministry of Economy, Trade and Industry (METI), that of Naraha, a small town affected by restrictions on residency after the 2011 Fukushima Daiichi nuclear accident, is an elucidatory case. The smart community management features a community energy management system enabling the mutual supply of surplus energy between commercial and disaster public housing facilities built after the 2015 lift of the residency ban and its management is entrusted to a local incorporated association,[16] Naraha Mirai, on behalf of the city government (Naraha Mirai, n.d.). At present, 40% of the total energy supply is locally generated by solar panels installed on buildings and its goal to become energy self-sufficient and carbon neutral by 2030. Similar projects have been initiated in nearby communities, such as Soma and Namie (Shigen enerugī shō, 2022).

To generalise, from the *jichitai* perspective, establishing a smart city model area proves one local administration's commitment to a certain set of values and ideas reflected in the central government's guidelines. At the local level, it serves as a locus where a common international framework is articulated locally, thus offering a cohesive leverage for local governments facing issues such as depopulation and slow economic growth, to increase their image and attract new residents.

Besides the energy domain, there are more than a hundred smart city projects broadly categorised as eco-cities,[17] transit-oriented development areas,[18] and resilient cities.[19] By enhancing ICT, particularly the Internet of Things, big data, vehicle automation, car sharing service (mobility as a service, MaaS), and biometrics-based services (such as automaltic remote bus-fare payments), each smart city project aims at the resolution of a specific issue. In addition, the Government of Japan has identified 30 "super cities" where intersecting issues are addressed in a more comprehensive manner, primarily through efficient data management (Prime Minister's Office of Japan, 2020).

On top of these initiatives, in the last decade, the Sustainable Development Goals (SDGs) framework has been actively promoted and widely adopted as a way to implement policy innovation and place identity and accountability on the part of the local governments. Thus, the Japanese government has designated more than 200 administrations as SDGs Model Cities[20] and Local Governments' SDGs Model Projects[21] (Cabinet Secretariat, Regional Revitalisation Bureau, 2020).

In light of the decentralisation trends described above, Tokyo's recent push on building a "Society 5.0", or a "society based on future technologies", where ICT and AI-based solutions (drones, robots, etc.) are implemented in various sectors such as agriculture, manufacturing, service and logistics to foster rural revitalisation (Tokunaga and Shimobe, 2021) or enhance local energy transition (Sokołowski, 2021a, 2021b) can be considered, in fact, as a part of the long-term plan to materialise decentralisation.

4. Mapping Japanese Smart Cities: Neoliberal Decentralisation at Work and the Unsolved Urban-Rural Divide

In order to examine the above-mentioned phenomena, we set out to compile a smart city project dataset based on available data highlighting the scale of individual projects. For the purpose of analysis, data from 468 projects, which were financed with national subsidies to achieve Society 5.0 between FY2017 and FY2023, were collected. As indicated above, the Government of Japan officially describes a smart city as a sort of experimental site for the materialisation of Society 5.0. In this sense, the listed projects can be considered Japanese smart city projects. The list and the number of projects are provided in Table 5.2.

Before proceeding to the analysis, we will briefly clarify the sampling method. First, the timeframe was set between fiscal years 2017 and 2023.[22] Fiscal year 2017 marked the first year that the idea of Society 5.0 started to be mentioned in the policy sphere of regional revitalisation, while the latter is the year for which the most recent data is available. Second, this list only includes the smart city projects that proceed by taking advantage of municipalities' legal responsibility for city planning. Although they are in the minority, some well-known Japanese smart city projects that are exclusively led by stakeholders from the private sector, such as Toyota's Woven City, have been omitted from the data for our analytical purposes.

Table 5.2 Smart city projects in Japan (2017–2023)

Ministry in charge	*Schemes*	*Sub schemes*	*Budget 2023 (2022)*	*2017*	*2018*	*2019*	*2020*	*2021*	*2022*	*2023*	*Total number of projects*
CAO	Digital Garden City Nation subsidy	Regional vitalization society 5.0	120,000 (100,000) million				13	6	8	2	29
	Implementation of future technology		93 (77) million		14	8	12	9	9	2	54
	National Strategic Zone	Super city	361 (301) million/ legal and taxation measure						2	0	2
		Digital health	361 (301) million/ legal and taxation measure						3	0	3
	Municipality SDGs*		704 (425) million		10	10	10	10	10	11	61
MIC	Smartcity to solve local issues		460 (460) million	6	3	5	7	9	13	8	51
MLIT	Japanese MaaS promotion prject		244 (142) million			20	35	11	6	7	79
	Smart city Model Project/smart city implementation project		370 (150) million			20	6	5	11	12	54
METI	Local MaaS promotion project		7400 (5850) million			13	16	14	11	8	62
MoEN	Local Decarbonization and new energy promotion subsidy		40,000 (20,000) million					25	20	28	73
GOJ	**Japanese Smart city Projects**			**6**	**27**	**76**	**99**	**89**	**93**	**78**	**468**

Note: * Among SDG Future City, only Municipality SDGs Model Projects are counted due to its financial support

In addition, due to its public nature, the data was publicly available from the official website of the cabinet and other relevant central ministries. All the subsidies listed on the Government of Japan's Smart City Public–Private Partnership Platform (PPPP) have been counted. These subsidies were launched under the late Prime Minister Abe's second tenure (2012–2020) within the policy framework of regional revitalisation. In addition to the projects listed on the PPPP, the list above also includes Municipality SDG Model Projects, which are flagship projects selected from the SDG Future City Scheme. These projects are important to consider because it contributed significantly to broadening the Japanese idea of smart cities from experimental projects of digitally informed advanced technologies to city planning with the ideas of sustainable development. Following this policy path, the Local Decarbonisation and New Energy Promotion Scheme subsidy was launched under the Suga cabinet (2020–2021), and more recently, Prime Minister Kishida (2021–2024) launched the DGCN Subsidy and National Strategic Zone. The amount of the Government of Japan's investment into smart cities has increased significantly over the years. While this pattern was certainly interrupted due to the global COVID-19 pandemic, it already shows signs of recovery.

The focus of our analysis is placed on the geographical scale categories of the municipalities that receive one or more of the above-listed subsidies. This scale is differentiated based on the *Urban Employment Area (UEA)*, a categorisation devised by economists Yoshitsugu Kanemoto and Kazuyuki Tokuoka in 2002. It captures the urban-suburban relationship by considering population concentration in terms of densely inhabited districts (DID) and individuals' commuting patterns. The UEA differentiates three scale categories based on the size of DID at its core, namely metropolitan, micropolitan, and the others. The metropolitan area includes one or more core cities with a DID population size exceeding 50,000. The micropolitan area has a DID population size less than 50,000 but more than 10,000 in its core city. Within metropolitan and micropolitan areas, there are core cities and suburbs, with the distinction identified based on individuals' commuting patterns for jobs and/or schooling. The other areas may also have DID, but their population size is limited to less than 10,000. In addition, these cities do not have any cities with inbound or outbound commuting patterns. This implies that these cities are often located in remote places, typically in mountainous areas or on islands. UEA was chosen for several advantages compared to other existing urban-suburban-rural categorisation methods. First, unlike population size, levels of depopulation, location, and/or official municipal categorisations provided by the Japanese government, UEA can dynamically identify rural and urban relationships based on people's commuting patterns. Second, even when compared to other dynamic categorisation methods, UEA can provide a more suitable economic-geographical configuration for Japan.

Kanemoto and Tokuoka (2002) explain that other existing geographical categories often employ the "one urban area and one urban core" principle, as they have been developed based on the American economic-geographical model. However, in the case of Japan, multiple urban cores form one gigantic metropolitan area and suburbs extend concentrically. This pattern has been constructed over time to enhance productivity by tightly weaving the web of economic activities within limited geographical spaces in Japan. The list of UEA is updated every five years and published online for public use (see Table 5.3).[23]

The locations of 468 smart cities have been identified and categorised into five different UEAs and an additional category labelled "Not Applicable (N/A)". This category refers to projects in which multiple municipalities are collaborating. It is important to note that the project percentage provided refers to the number of projects; this does not necessarily reflect the amount of financial support given. In fact, unlike previously existing transfer schemes – i.e. grant,[24] the entity of the subsidies[25] allocated by the central government to *jichitai* varies from project to project. In addition, due to the nature of the subsidy scheme, the granted amount may not necessarily match the amount of subsidy which will eventually be spent. Therefore, the closing report of the financial status from each project is essential. Calculating the financial distribution pattern remains a topic for future research.

First, Japanese smart city projects are undeniably urban-oriented. About 61.5% of all the smart cities are located in metropolitan areas; 35.5% of all the smart cities are located in the urban core of metropolitan areas. When the number of municipalities is considered, the urban core cities of metropolitan areas, accounting for 5.9% of the total number of municipalities, undertake 35.5% of the smart city projects. Per capita, the distribution seems less skewed; 48.2% of Japan's total population resides in the urban core cities in metropolitan areas, sharing 35.5% of the projects. However, all the listed projects, except for the National Strategic Zone for the Digital Health

Table 5.3 Scale of Japanese smart cities based on the UEA categories

Categories	*Subcategories*		*Number of projects*	*Percentage*	*Number of municipalities*	*Percentage*
Metropolitan Area	**Urban core**		**143**	36.7%	102	5.9%
	Suburban	Suburban 1	88		639	
		Suburban 2	8		124	
		Suburban 3	3		29	
		Suburban 4	0		2	
		Total	**99**	25.4%	794	46.2%
Micropolitan Area	**Urban core**		**21**	5.4%	97	5.6%
	Suburban	Suburban 1	25		204	
		Suburban 2	2		29	
		Suburban 3	2		3	
		Total	**29**	7.4%	236	13.7%
Other areas			**61**	15.6%	489	28.5%
Cross regional			**37**	9%		
Total			390	100%	1718	100.0%

Programme, have the core cities of metropolitan areas as their *de facto* main target. Indeed, all the municipalities in the urban core of the metropolitan areas receive one or more subsidies. Among the metropolitan areas, the Tokyo metropolitan area attracts the highest number of projects, with 62 projects. Of these, 38 projects are located in the core cities of the Tokyo metropolitan area, and 24 projects are located in its suburbs. The suburbs of Osaka and Nagoya follow in the rankings by attracting 21 and 20 projects, respectively. These projects similarly aim to enhance urban liveability through improved efficiency in managing energy, waste, mobility, and more, taking advantage of digitally informed advanced technologies. Local efforts often take the form of real estate development through infrastructure upgrades, legitimised within SDG 11: sustainable urban life. This result is consistent with not only what we have argued in the previous sections but also with what others have identified as a new stage of gentrification in both European and Asian contexts (Davidson and Lees, 2005; Shin and Kim, 2016). All in all, urban municipalities are more favoured to contribute to the national goal of achieving Society 5.0. It is rather ironic because the regional revitalisation policy, which is *de jure* the main politico-economic backbone of Japanese smart city projects, primarily aims at filling the existing regional gap between urban and rural areas.

Second, smaller municipalities are involved in severe competition for financial acquisition. About 12.8% of Japanese smart cities are located in micropolitan areas, which are home to 9.4% of the total population; 333 municipalities belong to micropolitan areas; however, only 60 smart city projects have taken place in this area. In addition, 15.0% of Japanese smart cities are located in the "other areas" category, which is home to 2.4% of the total national population; 70 smart city projects take place in 489 municipalities. Per capita, the micropolitan and other areas seem to be incentivised by the subsidy scheme. However, it is worth stressing that one municipality can receive multiple subsidies. This means that while some have been successful in securing state subsidies successively, the dominant majority of the municipalities in these categories have never been successful, applied, or even considered applying for the scheme. As we mentioned in the previous section, it is up to each municipality to decide whether to draft its city planning to benefit from the smart city scheme. The neoliberal aim to disentangle local administrations from their dependence on financial transfers from the central government is undoubtedly at work. All in all, urban municipalities seem to have benefited the most from the subsidy scheme for smart cities.

In order to contribute to a thicker description of the results of the above survey, we would like to briefly add an insight from field observations. One of the authors visited two municipalities of "the other" categories, which have been nominated as a model city for the municipality SDG scheme and the Local Decarbonisation and New Energy Promotion Scheme successively. Indeed, these municipalities, both from Okayama Prefecture, are rare winners of the subsidy distribution game in remote Japan. These municipalities have been critically concerned with the future of their community since the early 1990s. For both municipalities, the Heisei municipal merger during the 2000s has put enormous financial pressure on innovating their community management. Certainly, the pathways selected by the respective municipalities are different.

However, their efforts have been continuous ever since, employing key terminologies of regional policy at the national level, be it eco-cities (Fukuda administration), regional vitalisation and sustainability (Abe administration), decarbonisation (Suga administration), and the present digital and smart technologies (Kishida administration). Their engagement seems to have been fruitful; it resulted in human capital accumulation at the locality as well as empowerment in terms of national recognition of the locality as a light tower case. Over time, several municipalities have developed a strong politico-administrative pipeline to the central government and started to consult directly with officers at the national level, bypassing the conventional prefectural channel (hearings with municipal office, November 2022 and May 2023).

In this regard, these cases show a practical change, which accompanies an alternation of political-administrative structure. They may have materialised local vitalisation via community building projects towards Society 5.0 in a decentralised manner; however, what is important to remember is that they are two rare successful cases. Too much attention to successful cases may lead social scientists to look away from the overall political-administrative realities of decentralisation in Japan. The majority of regional revitalisation subsidies are still flowing into the municipalities of the metropolitan area, and many other municipalities remain to follow the conventional way of city governance.

5. Conclusion

It is possible to conclude that, in the Japanese context, smart city, as a policy idea, is contributing to the materialisation of the structural reforms initiated in the early 2000s. Smart cities are, in fact, the epitome of structural transformations in Japan's political economy and in the "sociotechnical network" underpinning it. Referring to the three stages of smart innovation elucidated above, it is possible to argue that a majority of Japan's smart cities are characterised by a mere infrastructural update (first stage). However, particularly after the 3/11, subsequent legislative interventions centred upon the neoliberal principles of decentralised governance have contributed, in some instances, to the rekindling of social roles within the processes of governance. This has inevitably led to a reduction of the central government's role and strengthened *jichitai* with regard to managing local affairs.

At the same time, as pointed out by Morozov and Bria (2018), the role of the private sector as a provider of financing and social infrastructure has clearly emerged. Nevertheless, in light of the above, it is clear that the smart city concept has become a way through which local governments represent themselves vis-à-vis the central government. Thus, when *jichitai* are able to successfully push forward smart city initiatives, they establish a mutually beneficial relation with the central government in so far as the latter is able to *rebrand* themselves keeping subsidies flowing in and, in exchange for that, the former can use the latter's experience and solutions for both domestic and foreign policy purposes (Zappa, 2020). This mechanism is however based on deliberate processes of inclusion and exclusion which, as pointed out above, favours metropolitan over micropolitan areas which should be, in principle, the targets of revitalisation initiatives. In fact, using Kanemoto and Tokuoka's (2002) UEA classification, a dominant majority of the municipalities in the micropolitan and other areas have been de facto excluded, due to lack of success or voluntary nonparticipation in the Government of Japan's subsidy schemes. The dominant neoliberal discourse serves as a means and justification of selection and concentration.

In this sense, despite being presented as poles of technological and social innovation, Japanese smart cities are seldom such. Rather, to quote Moser (2015), the smart city concept is deployed as "a new bottle for an old wine", or more poignantly, "glittering utopias" reproducing decade-old power dynamics and unevenness. In fact, we might argue that the smart city concept and smart city-related policies obliterate actual innovation for the sake of the central government's masterplans in the domains of regional revitalisation and infrastructural modernisation.

Nevertheless, in some cases, municipal offices consult directly with the central government, and towards the prefectural office, they simply "inform" what they have concluded with the central government. The relative political power balance between the municipality and the prefecture seems to have been modified. In these successful cases, the political power of local governments vis-à-vis the prefectural government has increased. Certainly, local governments in Japan still rely on financial transfers from the state. However, they have implemented structural changes through innovation in public service provision. The fiscal situation of these municipalities has been improving. These cases may point to a successful pathway to decentralised Japanese administration. Considering these facts, the political-administrative structure of the relation between centre and periphery has changed. In other words, inequality between urban and rural areas between large cities and small villages remains embedded in the present structure of the administrative relations between centre and periphery and depends on the pervasiveness of neoliberalism in the administration sector.

Notes

1 In Japanese: 持続可能な都市や地域 [*jizokukanō na toshi ya chiiki*].
2 In Japanese: Society 5.0の先行的な実現の場 [*Society 5.0 no senkōteki na jitsugen no ba*].
3 In Japanese: 未来技術社会実装 [*mirai gijutsu shakai jissō*].
4 In Japanese: 地方自治体 [*chihō jichitai*], or more simply 自治体 [*jichitai*].

5 In Japanese: エネルギーの使用の合理化に関する法律 [*enerugī no shiyō no gōrika ni kansuru hōritsu*].
6 In Japanese: 三位一体 [*sanmi ittai*].
7 In Japanese: 環境モデル都市 [*kankyō moderu toshi*].
8 In Japanese: まち・ひと・しごと創生法 [*machi-hito-shigoto sōsei hō*], Law No. 136/2014 repealed by the Act on the Establishment of the Digital Agency, Law No. 36/2021.
9 In Japanese: 総合戦略 [*sōgō senryaku*].
10 In Japanese: 地方版総合戦略 [*chihō ban sōgō senryaku*].
11 The term "Abenomics" refers to an economic strategy launched by the second Abe cabinet in 2013 in cooperation with the Bank of Japan aimed at spurring economic growth through the so-called three arrows: monetary easing, fiscal reform, and growth promotion (Schiff, 2015; Sokołowski, 2015). Hence, the expression "Local Abenomics" in Japanese ローカルアベノミクス [*rōkaru abenomikusu*] is attributed to the LDP MP Tatsuya Itō, chairman of the Research Commission, who used it to describe the government's initiatives to produce a "trickle-down" effect of Abenomics on regions. Here, the government should do the following: (1) support the creation of research and manufacturing clusters; (2) offer incentives to enterprises to foster ties with the local realities promoting a "virtuous cycle" benefitting the region's economy with an eye to foreign markets; and (3) encourage local authorities and communities to actively participate in the region's economic life (Itō, 2014).
12 In Japanese: 地方へ新しい人の流れを作る [*chihō e atarashii hito no nagare o tsukuru*].
13 In Japanese: 改善 [*kaizen*].
14 According to Samuels, in 2009, several SC projects were initiated under the Next Generation and Social Systems schemes launched by the Democratic Party of Japan-led government. The Yokohama Smart City project was built from scratch, while other flagship projects, such as Keihanna, Kitakyūshū, and Toyota were the continuation of previous subsidy schemes (such as the environmental model cities) (Samuels, 2013, p. 145).
15 In fact, smart communities may ensure reliable life continuity plans with regard to their capacity of securing energy in case of crisis (Okubo *et al.*, 2022).
16 In Japanese: 一般社団法人 [*ippan shadan hōjin*].
17 In Japanese: 環境共生都市 [*kankyō kyōsei toshi*].
18 In Japanese: 公共交通指向型都市 [*kōkyōkōtsū shikōgata toshi*].
19 In Japanese: 災害に強いまちづくり [*saigai ni tsuyoi machidzukuri*].
20 In Japanese: SDGs 未来都市 [*SDGs mirai toshi*].
21 In Japanese: 自治体SDGsモデル事業 [*jichitai SDGs moderu jigyō*].
22 The Japanese Fiscal Year begins on 1 April and ends on 31 March of the following solar year. Therefore, FY 2017 began 1 April 2017 and ended on 31 March 2018. FY 2023 began 1 April 2023 and ended 31 March 2024.
23 The data is available at: https://www.csis.u-tokyo.ac.jp/UEA/index_e.htm
24 In Japanese: 補助金 [*hojokin*].
25 In Japanese: 交付金 [*kōfukin*].

References

Baron, S., Patterson, A., Maull, R. and Warnaby, G. (2018) "Feed People First: A Service Ecosystem Perspective on Innovative Food Waste Reduction", *Journal of Service Research*, 21(1), pp. 135–150.

Barrett, B.F.D., DeWit, A. and Yarime, M. (2021) "Japanese Smart Cities and Communities: Integrating Technological and Institutional Innovation for Society 5.0", in H.M. Kim, S. Sabri and A. Kent (eds) *Smart Cities for Technological and Social Innovation*. London: Academic Press, pp. 73–94.

Cabinet Secretariat (2023) スマートシティガイドブック *2.0* [*Smart City Guidebook 2.0*]. Tokyo: Government of Japan. Available at: https://sbircao02-my.sharepoint.com/personal/kagisoukatsu1_sbircao02_onmicrosoft_com/_layouts/15/onedrive.aspx?id=%2Fpersonal%2Fkagisoukatsu1%5Fsbircao02%5Fonmicrosoft%5Fcom%2FDocuments%2Fcstp%2Fsociety5%5F0%2Fsmartcity%2Fsc%2Dguid%2Dbook%2D0%2D125%2D2%2Epdf&parent=%2Fpersonal%2Fkagisoukatsu1%5Fsbircao02%5Fonmicrosoft%5Fcom%2FDocuments%2Fcstp%2Fsociety5%5F0%2Fsmartcity&ga=1 (Accessed: 29 April 2024).

Cabinet Secretariat, Bureau for City, People, and Job Creation (2021) "地方版総合戦略の策定状況等に関する調査結果" [*Survey Results Regarding the Status of Formulation of Local Comprehensive Strategies*]. Available at: https://www.chisou.go.jp/sousei/about/chihouban/pdf/sakuteijoukyou210909.pdf (Accessed: 5 March 2023).

Cabinet Secretariat and Regional Revitalisation Bureau (2020) "「環境未来都市」構想・広域連携ＳＤＧｓモデル事業" [*'Future Eco-City' Initiative and Wide-Area Collaboration SDGs Model Project*]. Available at: https://www.chisou.go.jp/tiiki/kankyo/index.html (Accessed: 29 April 2024).

Davidson, M. and Lees, L. (2005) "New-Build 'Gentrification' and London's Riverside Renaissance", *Environment and Planning A: Economy and Space*, 37(7), pp. 1165–1190.

De Franco, A., Moroni, S. and De Lotto, R. (2023) "Energy Communities in Smart Urban Ecosystems: Institutional, Organisational, Psychological, Technological Issues", in M.M. Sokołowski and A. Visvizi (eds) *Routledge Handbook of Energy Communities and Smart Cities*. London: Routledge, pp. 13–26.

Deguchi, A. (2020) "From Smart City to Society 5.0", in Hitachi-UTokyo Laboratory (ed) *Society 5.0: A People-Centric Super-Smart Society*. Singapore: Springer, pp. 43–65.

DeWit, A. (2014) "Japan's Resilient, Decarbonizing and Democratic Smart Communities", *The Asia-Pacific Journal: Japan Focus*, 12(50/3). Available at: https://apjjf.org/2014/12/50/Andrew-DeWit/4236.html (Accessed: 2 May 2024).

Geels, F.W. (2002) "Technological Transitions as Evolutionary Reconfiguration Processes: A Multi-Level Perspective and a Case-Study", *Research Policy*, 31(8), pp. 1257–1274.

Gonella, F. (2019) "The Smart Narrative of a Smart City", *Frontiers in Sustainable Cities*, 1, p 9.

Government of Japan (2014) "まち・ひと・しごと創生法" [*Towns, People and Jobs Creation Act*]. Available at: https://elaws.e-gov.go.jp/document?lawid=426AC0000000136_20210901_503AC0000000036 (Accessed: 29 April 2024).

Greenfield, A. (2017) *Radical Technologies: The Design of Everyday Life*. London – New York, NY: Verso.

Hajer, M.A. (2005) "Coalitions, Practices, and Meaning in Environmental Politics: From Acid Rain to BSE", in D. Howarth and J. Torfing (eds) *Discourse Theory in European Politics*. London: Palgrave Macmillan, pp. 297–315.

Itō, T. (2014) "「ローカル・アベノミクス」を提言" [*Proposal for 'Local Abenomics'*], *Liberal & Democratic Jiminshū*, 24 June, p. 1.

Jichitai Tsūshin Online (2020) "まち・ひと・しごと創生法について、地方通信" [*Local News About the Towns, People and Job Creation Act*]. Available at: https://jt-tsushin.jp/articles/column/casestudy_machi-hito-shigoto_sosei (Accessed: 28 April 2024).

Kanemoto, Y. and Tokuoka, K. (2002) "日本の都市圏設定基準"[*Proposal for the Standards of Metropolitan Areas of Japan*], *Ōyō chiiki gaku kenkyū* [*Journal of Applied Regional Science*], 7, pp. 1–15.

Kimura, O. (2009) "The National Programs for Development of Energy Technologies", *SERC Discussion Paper*, SERC09007. Tokyo: Central Research Institute of Electric Power Industry, pp. 1–27. Available at: https://www.denken.or.jp/jp/serc/research_re/download/09007dp.pdf (Accessed: 1 March 2024).

Kishida, F. (2022) "岸田内閣総理大臣年頭記者会見" [*Prime Minister Kishida's New Year's Press Conference*]. Available at: https://www.kantei.go.jp/jp/101_kishida/statement/2022/0104nentou.html (Accessed: 30 April 2024).

Latour, B. (1990) "Technology is Society Made Durable", *The Sociological Review*, 38(1_suppl), pp. 103–131.

Morita, K., Okitasari, M. and Masuda, H. (2020) "Analysis of National and Local Governance Systems to Achieve the Sustainable Development Goals: Case Studies of Japan and Indonesia", *Sustainability Science*, 15(1), pp. 179–202.

Morozov, E. and Bria, F. (2018) *Rethinking the Smart City: Democratizing Urban Technology*. New York, NY: Rosa Luxembourg Stiftung, pp. 1–56. Available at: https://sachsen.rosalux.de/fileadmin/rls_uploads/pdfs/sonst_publikationen/rethinking_the_smart_city.pdf (Accessed: 20 November 2023).

Moser, S. (2015) "New Cities: Old Wine in New Bottles?", *Dialogues in Human Geography*, 5(1), pp. 31–35.

Nakazawa, H. (2014) "地方と中央 –「均衡ある発展」という建前の崩壊" [*The Centre and the Periphery: The End of the Myth of Equal Development*], in *Heisei shi*. Tokyo: Kawade shobō.

Naraha Mirai (n.d.) "私たちについて" [*About Us*]. Available at: https://narahamirai.com/aboutus/service/ (Accessed: 18 November 2022).

Niimura, T. (2018) "Decentralization Reform in Japan", *Seikei Hōgaku*, 89, pp. 95–108.

Okubo, H., Shimoda, Y., Kitagawa, Y., Gondokusuma, M.I.C., Sawamura, A. and Deto, K. (2022) "Smart Communities in Japan: Requirements and Simulation for Determining Index Values", *Journal of Urban Management*, 11, p. 500.

Prime Minister's Office of Japan (2020) "日本のスマートシティーSDGsなど世界が抱える課題を日本のSociety5.0で解決" [*Japan's Smart Cities: Resolving Global Issues Such as SDGs Through Japan's Society 5.0*]. Available at: https://www.kantei.go.jp/jp/singi/keikyou/pdf/smart_city_catalog.pdf (Accessed: 29 April 2024).

Sakuma, N., Trencher, G., Yarime, M. and Onuki, M. (2021) "A Comparison of Smart City Research and Practice in Sweden and Japan: Trends and Opportunities Identified From a Literature Review and Co-Occurrence Network Analysis", *Sustainability Science*, 16(6), pp. 1777–1796.

Samuels, R.J. (2013) *3.11: Disaster and Change in Japan*. Ithaca, NY: Cornell University Press.

Schiff, J.A. (2015) "Abenomics: From the Lost Decade to the Three Arrows", in D. Botman, S. Danninger and J. Schiff (eds) *Can Abenomics Succeed?* Washington, DC: International Monetary Fund. Available at: https://www.elibrary.imf.org/display/book/9781498324687/ch001.xml (Accessed: 1 March 2024).

Shigen enerugī shō (2022) "地域共生型再生可能エネルギー事例検証：事例集" [*The Local Mutualism for Renewable Energy Award: Case-Studies*], 2021 Ministry of Economy, Trade and Industry. Available at: https://www.enecho.meti.go.jp/category/saving_and_new/advanced_systems/saiene_kensho/doc/case-studies_r3.pdf (Accessed: 28 April 2024).

Shin, H.B. and Kim, S.-H. (2016) "The Developmental State, Speculative Urbanisation and the Politics of Displacement in Gentrifying Seoul", *Urban Studies*, 53(3), pp. 540–559.

Sokołowski, M.M. (2015) "Priorities of Energy Policy of Japan Under Abenomics", in M. Sitek and M. Łęski (eds) *Opportunities for Cooperation Between Europe and Asia*. Józefów: Alcide De Gasperi University of Euroregional Economy, pp. 227–240.

Sokołowski, M.M. (2021a) "Artificial Intelligence and Climate-Energy Policies of the EU and Japan", in D. Bielicki (ed) *Regulating Artificial Intelligence in Industry*. London: Routledge, pp. 138–155.
Sokołowski, M.M. (2021b) "Models of Energy Communities in Japan (Enekomi): Regulatory Solutions From the European Union (Rescoms and Citencoms)", *European Energy and Environmental Law Review*, 30(4), pp. 149–159.
Sokołowski, M.M. (2022) *Energy Transition of the Electricity Sectors in the European Union and Japan: Regulatory Models and Legislative Solutions*. Cham: Palgrave Macmillan.
Sokołowski, M.M. and Visvizi, A. (eds) (2023) *Routledge Handbook of Energy Communities and Smart Cities*. London: Routledge.
Sugiyama, N. and Takeuchi, T. (2008) "Local Policies for Climate Change in Japan", *The Journal of Environment & Development*, 17(4), pp. 424–441.
Tokunaga, T. and Shimobe J. (2021) "第1回 2021年度予算案から見たスマートシティの行方" [*Part 1: The Future of Smart Cities as Seen From the FY2021 Budget Proposal*]. Available at: https://project.nikkeibp.co.jp/atclppp/021900032/021900002/ (Accessed: 30 April 2022).
Trencher, G. (2019) "Towards the Smart City 2.0: Empirical Evidence of Using Smartness as a Tool for Tackling Social Challenges", *Technological Forecasting and Social Change*, 142, pp. 117–128.
Voorwinden, A. (2021) "The Privatised City: Technology and Public-Private Partnerships in the Smart City", *Law, Innovation and Technology*, 13(2), pp. 439–463.
Zappa, M. (2020) "Smart Energy for the World: The Rise of a Technonationalist Discourse in Japan in the Late 2000s", *International Quarterly for Asian Studies*, 51(1–2), pp. 193–222.

6 Learning the Characteristics of Vacant Houses

Smart Solutions for Japanese Municipalities

Yuki Akiyama

1. Introduction

In recent years, Japan has seen a continual rise in the number of vacant houses across the nation. This trend is attributed to a confluence of factors, including a declining birth rate, an ageing population, a resultant overall population decline, urban migration, and shifts in the industrial landscape. Figure 6.1, drawing on data from the latest Housing and Land Survey (2018) by the Statistics Bureau of the Ministry of Internal Affairs and Communications, illustrates the magnitude of this issue. According to the survey, the estimated number of vacant houses nationwide reached approximately 8.46 million, with a vacancy rate of 13.6% in 2018, marking the most recent period for which data is available. Moreover, both the number of vacant houses and the vacancy rate have shown an upward trend. Looking ahead, the situation appears poised to worsen; for instance, a study by Imai *et al.* (2015) presents a grim forecast, predicting that the number of vacant houses in Japan could escalate to 21.5 million by 2033, nearly 2.5 times the figure recorded in the 2018 survey.

1.1 What Is the Problem of Vacant Houses in Japan?

In the Housing and Land Statistics, vacant houses are delineated into three main categories based on their management status: (1) "Secondary housing": encompassing vacation homes and other dwellings used occasionally; (2) "Housing for rent or sale", and (3) "Other housing": houses not occupied due to reasons not mentioned in the first two categories. This third category, the other housing, includes homes left vacant due to the prolonged absence of occupants because of work-related relocations, hospitalisations, or those awaiting demolition for redevelopment. Such housing is often at a higher risk of poor management compared to the other categories. Figure 6.2, derived from the Housing and Land Survey (2018), illustrates the distribution of vacant house types across Japan, highlighting a notable increase in the proportion of the other housing. This trend of rising vacancies in the other housing category is expected to persist, with projections by Akiyama and Shibasaki (2015)

DOI: 10.4324/9781003471448-6

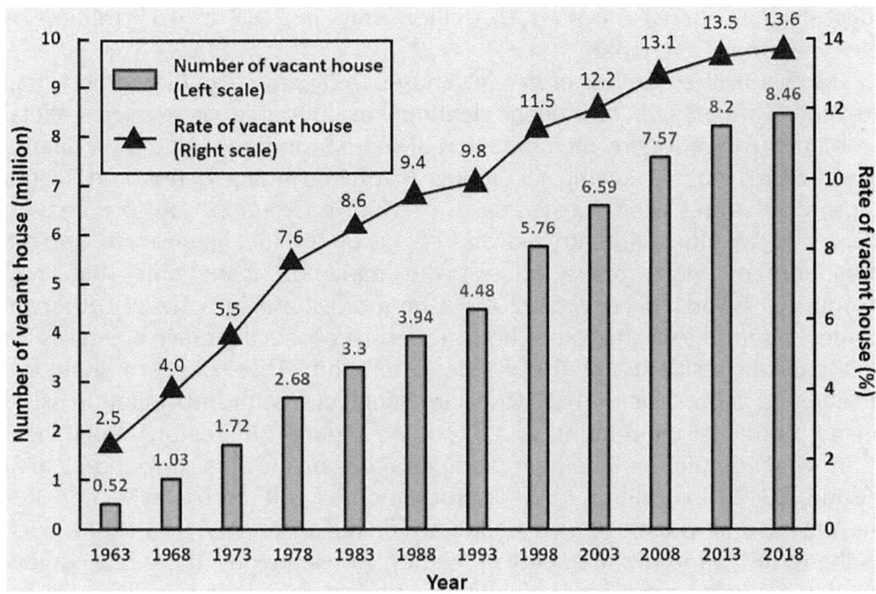

Figure 6.1 Number of vacant houses and vacancy rate across Japan
Source: (Housing and Land Survey, 2018)

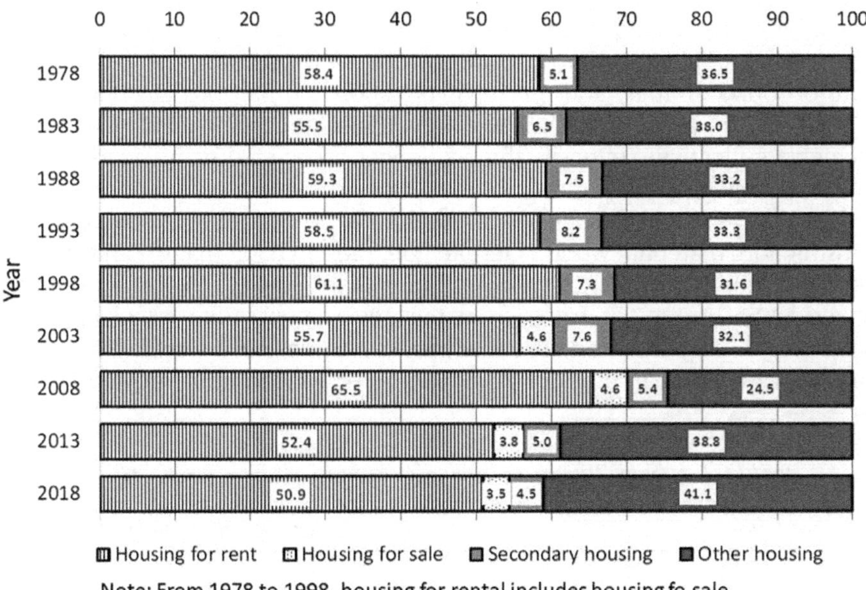

Figure 6.2 Trends in the percentage of vacant houses by type across Japan
Source: (Housing and Land Survey, 2018)

suggesting an increase from 3.18 million units in 2013 to 4.17 million by 2040 nationwide in Japan.

The rise in the number of vacant houses in Japan, particularly those that are poorly managed, has been identified as a source of several societal problems. The problems include external diseconomies such as the collapse of structures due to ageing, accidents involving falling debris and scatter caused by strong winds, a decrease in crime prevention due to increased arson and unauthorised entry, adverse effects on the local landscape, and the diminution of rental property values in the vicinity of vacant homes (Baba and Asami, 2017; Sadayuki *et al.*, 2020). Furthermore, a specific category of vacant homes, referred to as the other housing, poses especially grave concerns for neighbouring residents and the broader community. These concerns include a heightened risk of structural collapse due to neglect, health and sanitation issues arising from pest infestations, and in colder, snow-prone regions, the danger of snow accumulation leading to additional hazards for the surrounding area (Asami, 2014). In addition to these problems, as I will discuss in Section 1.3, the increase in vacant houses is directly linked to energy challenges, such as the reduction in urban energy efficiency. Consequently, there is an urgent need to explore strategies for effectively utilising, reclaiming, or demolishing these vacant properties to address energy concerns.

The phenomenon of increasing vacant houses is acknowledged as a social issue globally, particularly in developed countries (Bernt, 2009; Silverman *et al.*, 2013; Couch and Cocks, 2013). In Japan, this issue is particularly exacerbated by a unique combination of factors: the world's fastest-ageing population, significant urban migration of the youth pursuing higher education and job opportunities, and the relatively short lifespan of detached homes (Komatsu and Endo, 2000; Tanikawa and Hashimoto, 2009) (Table 6.1). These elements collectively intensify the challenges associated with vacant houses in Japan, making the situation more severe than in other countries.

The relatively short lifespan of Japanese houses is not due to their predominant construction material, wood. Despite the presence of many centuries-old wooden buildings in Japan, such as shrines and temples, the brief life cycle of residential properties stems from the government's real estate appraisal practices. These practices dictate that the asset value of a house depreciates to zero after 20 years. In the post-World War II era, Japan implemented various

Table 6.1 Average life expectancy of houses by country (MLIT, 2006)

Nation	*Average life span (year)*
Japan	30
Germany	79
France	86
United States	103
United Kingdom	141

incentives and subsidies for new constructions as part of post-war reconstruction efforts, promoting the idea that new building projects would drive the nation's industrial growth. However, this approach led to the depreciation of older homes' value, with the land's value being the only remaining asset over time. This policy has been a significant factor contributing to the short lifespan of housing in Japan. Recent efforts in Japan have begun to focus on extending the lifespan of residential buildings (Minami, 2021), yet these efforts are only partially successful. Many homes still feature poor insulation and continue to use outdated, inefficient heating, cooling, and lighting systems, resulting in low energy efficiency. It has been noted that such properties tend to decrease in residential value and are more likely to become vacant (Sasatani *et al.*, 2010).

1.2 Measures for Vacant Houses and Its Problems in Japan

As highlighted earlier, the management of vacant houses is becoming increasingly important. In response, the "Vacant Houses Special Measures Act" was implemented in May 2015. Before this Act, local governments attempted to address the issue using ordinances that encouraged the utilisation and removal of vacant houses. However, these measures lacked legal authority, leaving the final decisions regarding vacant properties in the hands of their owners and, consequently, proving to be insufficiently effective. With the introduction of this Act, houses deemed to be poorly managed can now be classified as the "specified vacant houses" following assessments by local authorities. This classification enables local governments to provide guidance to owners and advocate for necessary improvements. In a significant update, the Act was revised in December 2023. This revision introduced the category of the "unmanaged vacant houses" as potential candidates for becoming the specified vacant houses and authorised the government to take pre-emptive actions. These actions include offering guidance and recommendations for improvements before a property is officially designated as a specified vacant house (Act No. 50 of 2023).

For local governments to address the issue of vacant houses effectively and efficiently, it is imperative first to understand the spatial distribution of vacant houses within their municipalities. While numerous studies in Japan have aimed to map out the spatial distribution of vacant houses (Mashita and Akiyama, 2020), the majority have relied on methods such as individual visual inspections through field surveys (Mizusawa *et al.*, 2021; Asano and Inoue, 2022, etc.) or interviews and questionnaires targeting local residents (Shinobe and Urabe, 2014; Sumikura *et al.*, 2020, etc.). In addition, the approach taken by local governments to survey vacant houses often involves visual inspections of each house's exterior through individual visits, making the current methods of surveying by local authorities time-consuming, costly, and labour intensive (Administrative Evaluation Bureau *et al.*, 2019). This extensive requirement of resources has posed a significant problem for municipalities in implementing vacant house measures across broad areas.

1.3 Vacant House Problem from the Perspective of Energy Issues

Vacant houses significantly impact urban energy issues in several ways. The "specified vacant houses" referred to earlier often consist of older buildings with diminished functionality as residences. These deteriorating structures exhibit poor energy efficiency, largely due to inadequate insulation and the absence of modern energy-saving appliances such as LED lighting, high-efficiency heating and cooling systems, and water-efficient heaters that have become standard in recent constructions. It has been highlighted that leaving these properties unattended leads to unnecessary energy consumption through electricity and water leakages (Jia *et al.*, 2012). Moreover, the unchecked increase in vacant houses within a city hampers the initiation of large-scale redevelopment projects, as these structures need to be cleared away first. This situation poses a barrier to transforming the area into an energy-efficient urban space with a high quality of life. Consequently, this problem has been known to skew the supply-demand balance in the real estate market, driving up housing prices (Sun *et al.*, 2011).

On the contrary, vacant houses hold the potential to aid in addressing energy challenges, contingent upon their tilization. For instance, when situated in commercial zones, these properties can be repurposed into retail spaces (Joo *et al.*, 2022), thereby mitigating issues of neglect and promoting active oversight by local authorities. Moreover, vacant properties and their surrounding land can serve as venues for renewable energy projects, significantly contributing to energy solutions. Notably, the installation of solar panels on the roofs and lands of vacant houses for electricity generation has been recognised for its role in creating numerous small-scale power generation sites within cities. These sites are strategically placed near urban areas where electricity demand is high, thereby reducing transmission losses (Ignacio and Soria, 2021). This approach has also shown promise in rural settings (Pratheeba *et al.*, 2023). Furthermore, it has been observed that well-maintained vacant houses, which meet the standards for reoccupation, can be transformed into energy-efficient homes through renovations. By incorporating insulation, double-glazed windows, and other energy-saving modifications (Deng *et al.*, 2023), these houses can become a source of affordable, energy-efficient, and habitable housing options, offering a cost-effective alternative to new constructions.

The initiatives discussed are crucial in the context of advancing smart cities. Tokyo, the world's largest megacity, is also among the most energy-efficient cities globally (Aki *et al.*, 2022). In addition, Tokyo has set an ambitious goal of achieving net-zero CO_2 emissions by 2050 (Yamada *et al.*, 2023). To meet this significant target, it is essential to substantially increase the energy efficiency of housing, including vacant houses. If Tokyo's strategies are thoroughly developed and actively promoted, they could serve as a model for other cities and regions in Japan, transforming them into "smart energy cities". Since the oil shocks of the 1970s, Japan has maintained a commendable international reputation in the energy sector, with concerted efforts from both public and private sectors to enhance energy efficiency and compensate for its scarcity of natural resources (Holroyd, 2018). Although its reputation was temporarily marred by the nuclear accident following the 2011 Great East Japan Earthquake, Japan's positive image in the energy sector remains robust. Consequently, Japan's endeavours to establish smart energy cities could potentially serve as exemplary models for other nations globally.

As discussed above, the issue of vacant houses is intricately linked to urban energy challenges and addressing the former can significantly contribute to resolving the latter. Thus, identifying the distribution of vacant houses represents a critical initial step in tackling the vacant house issues and is an essential strategy in addressing urban energy issues.

2. Previous Research Efforts

In response to the vacant house issues outlined above, the authors have developed a method for surveying the spatial distribution of vacant houses within municipalities quickly and cost-effectively, utilising a variety of data already held by the municipalities (hereafter referred to as "municipal data"). The distinctiveness of this approach lies in its use of previous surveys on vacant house distributions at a part of their municipality as a form of training data, which, when integrated with various municipal and private company datasets, allows for an estimation of the spatial distribution of vacant houses across the entire municipality. Municipal data encompasses both "open data" – data that can be freely used for secondary purposes such as research and analysis by external entities – and "closed data" – data not publicly accessible and is used internally within the municipality. The motivation behind employing these datasets is to significantly reduce the costs associated with conducting vacant house distribution surveys for local governments by leveraging the data that all Japanese municipalities already possess, maintain, and regularly update.

This initiative serves as a prime example of smart solutions for local governments in Japan. By employing AI to analyse data assets already in possession of local governments, this project offers a solution to the vacant house issue with a speed and cost efficiency previously unattainable through traditional municipal methods. In Japan, digital transformation (DX) is being actively promoted as part of smart city promotion efforts, with the aim of improving the efficiency of municipal operations using digital technology (Mitsunaga *et al.*, 2021; Ueda, 2023). Smart solutions in the existing research are expected to promote DX in municipalities and contribute to the active use of administrative data for the realisation of smart cities, solving the problem of vacant houses, and improving the energy efficiency of entire cities. These efforts can then be expected to contribute to the realisation of smart energy cities in the future.

Section 3 introduces a study by Akiyama *et al.* (2020, 2021) that estimates the distribution of vacant houses leveraging municipal data. This section has been thoroughly revised and updated from the original work by Akiyama *et al.* (2021). The focus of the next section is on detached houses, as it has been noted that in Japan, the methods for identifying vacant house distributions differ significantly between detached houses and apartment buildings (Ueda *et al.*, 2016). Furthermore, considering the aforementioned demand to eliminate external diseconomies caused by vacant houses, to understand the distribution of vacant houses that could become specific vacant houses, and to improve the energy efficiency of the entire city, it is considered more reasonable to target detached houses.

3. Example of Vacant House Distribution Estimation Using Municipal Data

This section presents a study by Akiyama *et al.* (2021) that focuses on estimating the distribution of vacant houses in Kagoshima (city), Kagoshima Prefecture, and Asakura (city), Fukuoka Prefecture.

3.1 Target Area

As previously introduced, the target area of this study was Kagoshima and Asakura cities. I secured cooperation from the local governments of these municipalities for conducting field surveys and accessing municipal data. Kagoshima, in particular, serves as a representative microcosm of Japanese municipalities, with its ageing rate, average household size, average household income, and vacancy rate all aligning closely with national averages. Furthermore, these municipalities encompass diverse areas, including residential zones, mixed-use areas combining residential, commercial, and industrial spaces, and regions characterised by both flat and sloping terrains. The rationale for choosing areas with such varied regional characteristics, as Akiyama *et al.* (2018) suggested, is to consider the influence that geographical traits may have on the prevalence of vacant houses. It is for these reasons that Kagoshima and Asakura were selected as the study's target areas.

3.2 Field Survey

To comprehend and estimate the spatial distribution of vacant houses using data from municipal data sources, it is crucial to identify the characteristics of vacant and non-vacant houses based on these data. This necessitates acquiring information on a certain number of buildings that are vacant. Therefore, in the study areas, I set up sample field survey areas to collect information on a certain number of vacant buildings, and conducted door-to-door visual field surveys in these areas to collect information on buildings that are actually vacant. Figure 6.3 illustrates the study area and the field survey areas of both cities. The study area was segmented into a 500-m grid that encompassed the entire city area, focusing on all grids that contained at least one detached house. Following discussions with local governments, which granted a permission to conduct the field surveys, I strategically chose survey areas to encompass a balanced mix of the central city area, residential areas near the

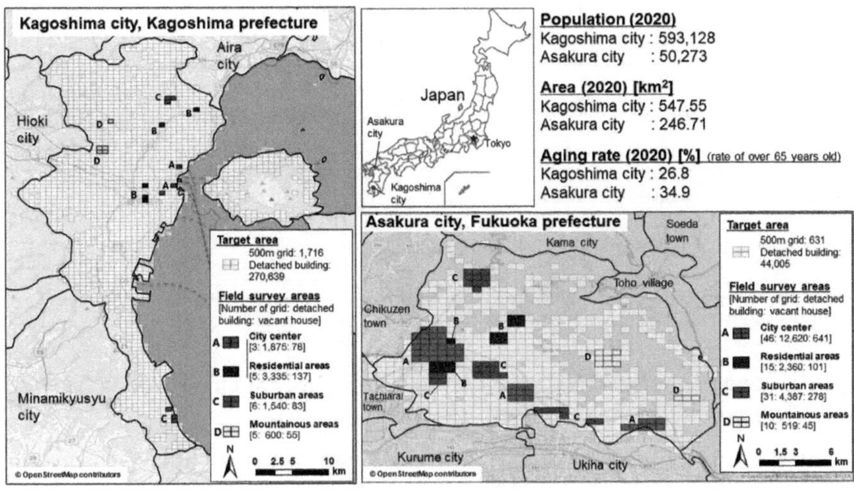

Figure 6.3 Study area and field survey area (Kagoshima and Asakura)

Figure 6.4 Field survey and examples of detached houses determined to be vacant (Kagoshima city)

city centre, suburban regions, and mountainous areas. Figure 6.4 displays the field survey and examples of buildings identified as vacant houses, while Table 6.2 provides the date of the field survey, the total number of buildings examined, and the tally of vacant houses documented from the field survey.

Table 6.2 Date of field survey, number of surveyed buildings, and number of surveyed vacant houses obtained from the field survey

City name	*Date of field survey*	*Number of surveyed houses*	*Number of surveyed vacancies*	*Rate of vacancy (%)*
Kagoshima	August 22–26, 2016	7,350	353	4.80
Asakura	October 17–21, 2016	19,886	1,065	5.36

3.3 Development of the "Vacant House Database" for Estimating the Distribution of Vacant Houses

By integrating the spatial distribution of vacant houses gathered from field surveys, with specific information on buildings derived from municipal data, and a digital residential map (a comprehensive digital map featuring building polygon data for all of Japan, provided by Zenrin Co., hereafter referred to as "Residential map"), the "Vacant House Database" (hereafter referred to as "VHD") was created. VHD serves as a tool for analysing the characteristics of vacant houses. Table 6.3 outlines the attributes of the VHD. From the "(1) Residential maps", I obtained building use and area. Residential maps are considered private data; they are extensively utilised by numerous municipalities across Japan, and the ease of access to these maps for municipalities is relatively high, hence their use in this study. The data from the field survey were employed under the category "(2) Field Survey". The "(3) Basic Resident Register (BRR)", "(4) Hydrant Consumption Amount Information (HCI)", and "(5) Building Registration Information (BRI)" represent the municipal data utilised in this research, all of which were procured from local governments.

The basic resident register (BRR) (3) is a comprehensive database of all residents within a municipality. The hydrant consumption amount information (HCI) (4) is maintained by the municipal water department and details water consumption by hydrants. HCI allows for the assessment of water usage by individual taps and the identification of inactive taps. Building registration information (BRI) (5) includes data on individual plots of land and buildings, maintained by local governments and legal affairs bureaus; this study focuses specifically on the information pertaining to buildings. Numerical land information (6) refers to GIS data publicly available from the Ministry of Land, Infrastructure, Transport and Tourism (MLIT), classified as open data. For this study, designated land use zones derived from urban planning summary maps prepared by local governments across Japan were used.

To integrate the aforementioned datasets (2) through (6) with the residential map (1), quantitative locational information, namely coordinates in

Table 6.3 List of attributes of the Vacant House Database (VHD)

Data source	*Survey date*	*Attribute*	*Data type*
(1) Residential map	K: 2016	Building ID	V
	A: 2016	Building use	C
		Building area (m^2)	V
(2) Field survey	K: August 22–26, 2016	Field survey area name (In case of field survey area)	C
	A: October 17–21, 2016	Evaluation result of vacant house	C
(3) Basic resident register (BRR)	K: June 2016 A: September 2016	Evaluation result of whether BRR could be integrated	C
		Number of households	V
		Number of residents	V
		Average age of resident	V
		Age of the youngest resident	V
		Age of the oldest resident	V
		Shortest residential household year	V
		Longest residential household year	V
(4) Hydrant consumption amount information (HCI)	K: June 2016 A: October 2016	Evaluation result of whether HCI could be integrated (non-integrated, integrated and opened, integrated and closed)	C
		Duration of hydrant close from last closed year (year)	V
		Annual consumption amount (*t*)	V
(5) Building registration information (BRI)	K: January 2016 A: January 2016	Evaluation result of whether BRI could be integrated	C
		Building age (year)	V
		Building structure	C
		Building use	C
(6) National land numerical information	K: 2011 A: 2011	Designated land use zone	C

Note: "V" in the "Data type" means numerical data, "C" means categorical data; "K" in the "Survey date" means Kagoshima (city); and "A" means Asakura (city).

longitude and latitude, is required. Since (2) includes coordinate information generated through field surveys and (6) consists of GIS data (in shapefile format), they can be directly spatially integrated with (1) using GIS technology. Conversely, datasets (3), (4), and (5), which lack coordinate information but include address details, can be converted into coordinate information through address matching. As a result, datasets (3), (4), and (5) can also be spatially integrated with (1).

The datasets (3), (4), and (5) constitute municipal data, intended for use within the respective municipality and, as a principle, not accessible to individuals outside the municipality. For this study, permission for research utilisation was granted by the Personal Information Protection Review Board of both cities, which are the study's target areas. Consequently, the data were handled confidentially to ensure that the authors did not directly access personal information, such as the names of residents.

It is important to recognise that not every building represented on residential maps can be successfully linked to municipal data, due to several factors. Some challenges were encountered with address notations being incomplete or the address data for matching purposes being unreliable, leading to a low accuracy in address geocoding. In addition, in the context of HCI, certain houses, particularly those in mountainous areas that rely on well water, could not be linked to HCI. Moreover, the rate at which municipal data was linked to houses identified as vacant through field surveys was found to be lower compared to that for houses not considered vacant. This finding indicates the utility of municipal data in discerning the vacancy status of a house. However, it is important to acknowledge the need for developing a model capable of evaluating the significance of missing municipal data, as the absence of data linkage itself can offer meaningful insights.

3.4 Development of a Machine Learning Model for Estimating Building Vacancy Probabilities

In this study, I utilised XGBoost (eXtreme Gradient Boosting), a decision tree-based machine learning model, to estimate the vacancy probability of individual houses. Introduced by Chen and Guestrin (2016), XGBoost falls under the category of gradient boosting methods. Gradient boosting is an ensemble learning method that incrementally adds weak learners (in this case, decision trees) to enhance the model's predictive accuracy. This sequential addition of new decision trees to the model helps in reducing errors and enhancing its generalizability. While the intricate computational details are documented by Chen and Guestrin (2016), employing this method enabled us to accurately estimate the vacancy probabilities of houses, effectively incorporating and adapting to previously unknown information, as discussed in this study.

There are several representative decision tree-based models, including Random Forest, LightGBM (Light Gradient Boosting Machine), and XGBoost. The decision to prefer XGBoost over other methods in this analysis was made for several reasons. While Random Forest has the capability to compute each decision tree in parallel, facilitating rapid estimations, its accuracy often falls short of XGBoost, which prioritises minimising estimation errors. LightGBM, introduced in 2016, is a relatively newer method. Although its estimation accuracy is comparable to or slightly lesser than that of XGBoost, depending on the dataset used (Liang *et al.*, 2020), it boasts a lower computational cost. However, given the sample sizes in this analysis, ranging from tens to hundreds of thousands, computation time was not deemed a critical factor. Furthermore, XGBoost, predating LightGBM, has been applied successfully in solving similar problems, such as estimating building heating loads (Le *et al.*, 2019), identifying prospective rental housing tenants (So and Arai, 2020), and forecasting housing prices (Chen *et al.*, 2023). This history suggests XGBoost's reliability and appropriateness for future use by local governments. For these reasons, XGBoost was the chosen method.

In developing the XGBoost model, it is essential to partition the data from the entire study area into two sets: training data for model development and evaluation data for tuning the model's parameters. Consistent with many existing studies, I allocated 70% of the data to training and 30% to evaluation. The explanatory variables used in this study are detailed in Table 6.4. Given that building use, building structure, and zoning from residential maps and BRI are categorical data, they were converted into dummy variables, categorised into 5, 4, 5, and 14 types, respectively. Consequently, a total of 39 explanatory variables were incorporated into the analysis. Parameter tuning was conducted through grid search, as XGBoost necessitates the specification of various parameters. Specifically, parameters for the maximum tree depth, the minimum weighted sum of data required at child nodes, the percentage of columns to be randomly sampled at each tree, and the rate of random sampling at each tree were optimised using the cross-validation method. The default learning rate is set at 0.03 but was adjusted to 0.01 to mitigate overfitting. It is important to note that the model was individually constructed for each of the two cities to ensure high accuracy. This method enabled the estimation of not only the vacancy probability for each building but also the total number of vacant houses and the vacancy rate for each defined spatial unit, such as a grid.

Table 6.4 Accuracy of varying the thresholds for vacant house rate estimated for vacant house in the field study area (Kagoshima)

Threshold	*ENV*	*REC (%)*	*ENnV*	*SPE (%)*	*ACC (%)*
0.1	322	91.21	6,393	91.37	91.36
0.2	311	88.10	6,598	94.30	94.00
0.3	300	84.99	6,727	96.14	95.61
0.4	291	82.44	6,798	97.16	96.45
0.5	288	81.59	6,841	97.77	96.99
0.6	283	80.17	6,889	98.46	97.58
0.7	280	79.32	6,909	98.74	97.81
0.8	270	76.49	6,939	99.17	98.08
0.9	250	70.82	6,959	99.46	98.08

Note: "ENV"/"ENnV" is the number of houses estimated to be vacant/non-vacant according to the field survey. "REC"/"SPE" is the recall (ENV/Number of vacant houses by field survey) and the specificity (ENnV/Number of non-vacant houses by the field survey). "ACC" is the accuracy ((ENV + ENnV)/Number of houses by field survey).

3.5 Results of Vacant House Distribution Estimation

In this approach, each building is assigned a vacancy probability. Consequently, buildings with a vacancy probability exceeding a certain threshold are classified as "vacant house", while those below this threshold are deemed "non-vacant house" Tables 6.4 and 6.5 present the accuracy results of applying various threshold values to distinguish between vacant and non-vacant statuses based on estimated vacancy probabilities in the field survey areas of the target two cities. These findings reveal that the accuracy of identifying a building as vacant, when its vacancy probability surpasses a predetermined threshold, relies on several metrics: (1) the number of houses accurately identified as vacant in the field survey (ENV); (2) the proportion of correctly identified vacant buildings (recall ratio, REC); (3) the proportion of buildings correctly identified as non-vacant in the field survey (ENnV); (4) the proportion of correctly identified non-vacant buildings (specificity, SPE); and (5) the overall accuracy of predictions matching the actual field survey outcomes (accuracy rate, ACC).

3.5.1 Verification of Model Accuracy

In both cities, as the threshold value increased, REC decreased while SPE increased. It was observed that a threshold of approximately 0.1 enabled the identification of vacant houses with an accuracy exceeding 90% in both cities. Conversely, a threshold of 0.7 or higher resulted in an accuracy rate of 97% or more in both locales. However, as illustrated in Figure 6.5, increasing the threshold value to enhance the precision of identifying non-vacant houses (SPE) inversely impacts the accuracy of identifying vacant buildings (REC). Figure 6.5 demonstrates that a threshold value of 0.7 marks a turning point in the trade-off between SPE and REC across varying threshold levels in both cities. Thus, setting the threshold at 0.7 optimises both SPE and REC values, maintaining high accuracy in classifying buildings as vacant or non-vacant.

Table 6.5 Accuracy of varying the thresholds for vacant house rate estimated to be vacant house in the field study area (Asakura)

Threshold	*ENV*	*REC (%)*	*ENnV*	*SPE (%)*	*ACC (%)*
0.1	982	92.21	13,441	71.41	72.53
0.2	941	88.36	15,482	82.26	82.59
0.3	908	85.26	16,774	89.12	88.92
0.4	877	82.35	17,567	93.34	92.75
0.5	859	80.66	18,036	95.83	95.02
0.6	844	79.25	18,319	97.33	96.36
0.7	824	77.37	18,515	98.37	97.25
0.8	709	66.57	18,665	99.17	97.43
0.9	508	47.70	18,764	99.70	96.91

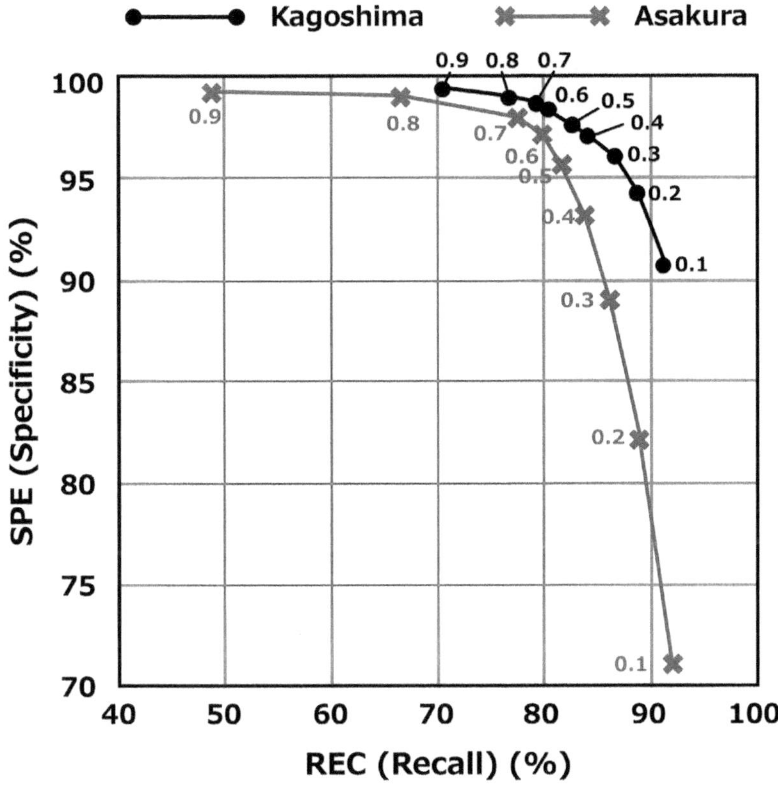

Figure 6.5 Relationship between REC and SPE of each threshold

Figure 6.6 illustrates the comparison between the estimated and actual values of vacant house determinations for each building within a specific area of Kagoshima, using a threshold value of 0.7. The results demonstrate that the estimations made by a method closely align with the actual values for nearly all buildings, whether vacant or non-vacant. Although not included in this chapter due to space constraints, it was verified that the estimation results and actual values of vacant buildings in most other surveyed areas also showed a high degree of agreement, as depicted in Figure 6.6.

Consequently, after assigning each building a vacant or non-vacant status using a threshold of 0.7, I aggregated the data using a 250-m grid to compare the actual and estimated number of vacant houses per grid. The results of this comparison are presented in Table 6.6. In both cities, a remarkably strong correlation was observed, with an *R*-value greater than 0.9. The mean absolute error (MAE) is also less than 1 in both cities, indicating that the discrepancy between the estimated and actual numbers of vacant houses per mesh averages less than one. These findings suggest that the method not only correlates well with the actual values across the entire city, as evidenced

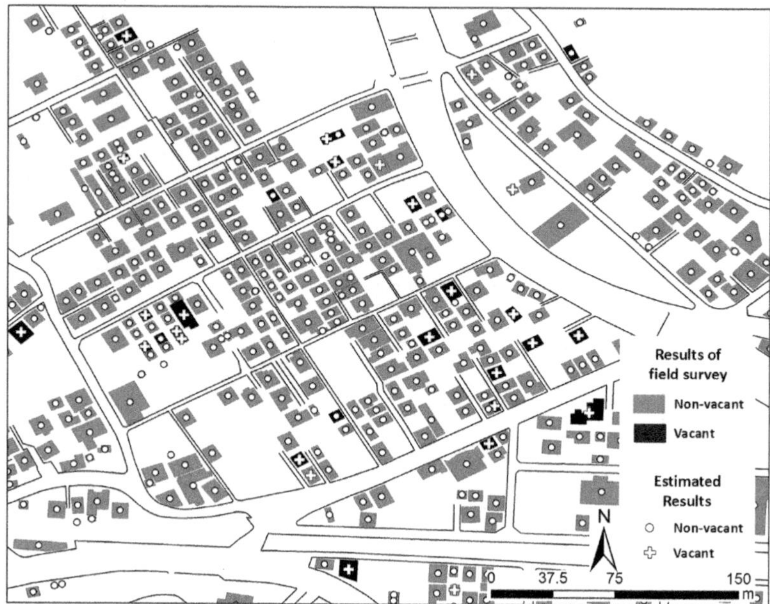

Figure 6.6 The results of comparing the vacant house estimation for each building with the true values from the field survey in a certain area of Kagoshima

Table 6.6 Reliability of estimated number of vacant houses by the method of this study (when the threshold is 0.7 and vacant houses are estimated for each building)

	Kagoshima	*Asakura*
R	0.9691	0.9476
AR^2	0.9392	0.8979
MAE	0.8054	0.6766
RMSE	1.1681	1.3202
p-value	3.33e-41	9.21e-35

in Tables 6.4 and 6.5, but also accurately estimates the spatial distribution of vacant houses with considerable reliability and without significant bias. Furthermore, these results affirm the utility of the XGBoost machine learning model in fulfilling the study's objectives and highlight the high extrapolative capability of this method.

3.5.2 Results of Citywide Vacant House Distribution Estimates

Figures 6.7 and 6.8 depict the estimated number of vacant buildings and the estimated vacancy rate per 500-m grid across the entirety of both cities. These

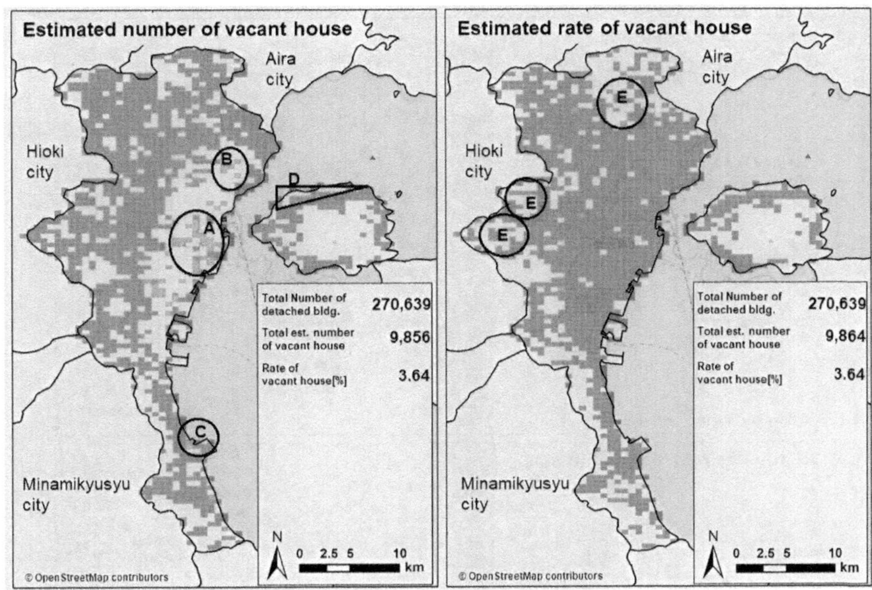

Figure 6.7 Estimated number of vacant houses (left) and estimated vacancy rate (right) by 500-m grid for the entire Kagoshima

results were derived by setting the threshold to 0.7, as indicated in Tables 6.4 and 6.5, to identify vacant houses on an individual building basis, followed by aggregating these findings for each 500-m grid unit. In Kagoshima, the total estimated number of vacant houses was 9,856, with a notably high concentration of vacancies in and around the central city area, particularly near City Hall (Figure 6.7, left A). A significant number of vacant houses were also observed in the newly developed residential zones on the hills surrounding the central city area (Figure 6.7, left B). In addition, areas with a high number of vacant houses and vacancy rates included the vicinity of the former town hall of the municipality before its amalgamation into Kagoshima (Figure 6.7, right C: former Kiire town urban area), the region around Sakurajima, an active volcano (Figure 6.7, right D), and the mountainous western part of the city. It was further noted that areas with particularly elevated vacancy rates are extensively spread throughout the mountainous regions in the western part of the city (Figure 6.7, right E).

In the case of Asakura, the total estimated number of vacant houses was 2,219. A notably high concentration of vacant houses was observed in the central city area around the city hall (Figure 6.8, upper A) and in the vicinity of the former town hall prior to its merger with Asakura (Figure 6.8, upper B and C). In addition, areas characterised by a high percentage of vacant houses were primarily found in the mid-mountainous regions (around Figure 6.8, lower D). This pattern mirrors that of Kagoshima, indicating that the distribution

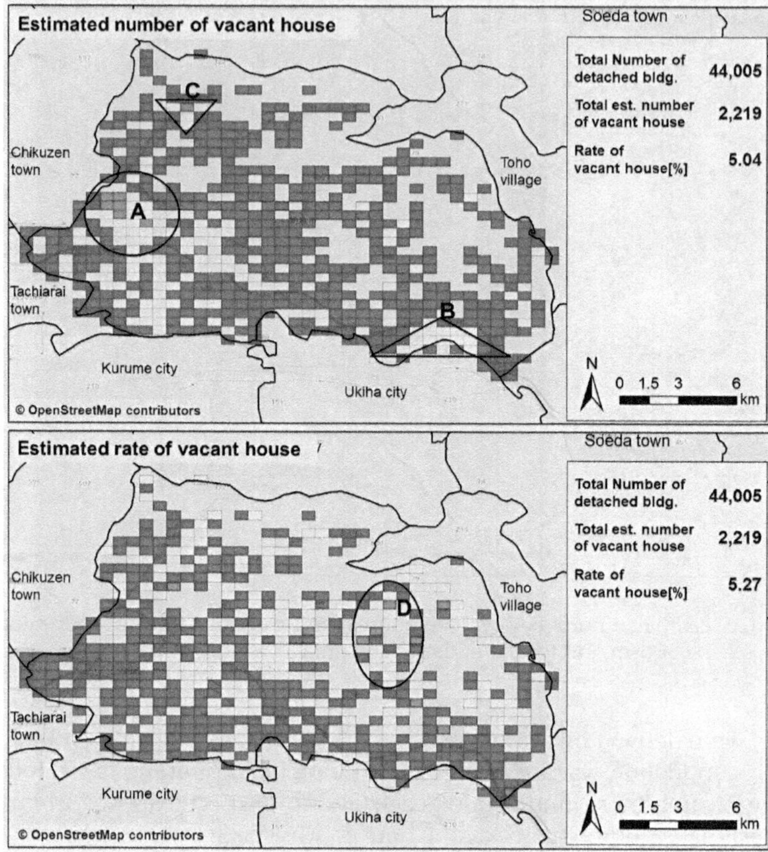

Figure 6.8 Estimated number of vacant houses (top) and estimated vacancy rate (bottom) by 500m grid for the entire Asakura City

trend of vacant houses tends to be consistent across municipalities, regardless of their size and the specific locations of these distributions.

By compiling the individual building estimation results into a grid as outlined, it becomes feasible to ascertain the spatial distribution of vacant houses from an area-wide perspective. Aggregating the data in this manner allows for the identification of specific areas within the city that exhibit notably high numbers of vacant houses or elevated vacancy rates. Such information is invaluable for local governments in determining which areas should be given priority in the implementation of vacant house mitigation strategies.

3.6 Proposals for Methods to Reduce the Burden of Field Surveys

Considering the practical application by local governments, the capability to pinpoint the distribution of vacant houses on an individual basis is highly valued. However, many Japanese local governments have a consensus that the final verdict on whether a building is vacant should be made through a field survey. Interviews with several local governments in Japan by the authors revealed that making a vacancy determination based solely on data, such as that presented in this study, is challenging due to the risk of overlooking genuinely vacant buildings. In other words, it was suggested that the ultimate decision should be reached via field survey.

Conversely, if the findings of this study can be leveraged to preliminarily identify buildings with a high likelihood of being non-vacant, then it allows for the classification of the rest as "suspected vacant buildings". Consequently, the use of this study's results can significantly reduce the number of buildings requiring field surveys, a departure from the previous necessity to inspect every building within the target area. As indicated in Tables 6.4 and 6.5, the accuracy in identifying non-vacant houses (SPE) can be enhanced by adopting a higher threshold value. For instance, in Kagoshima, setting the threshold at 0.9 ensures that 99.49% of houses classified as non-vacant are indeed not vacant houses. Hence, buildings flagged as non-vacant by the method are almost certainly non-vacant, and only those not identified as such warrant further field survey as potentially vacant buildings.

By applying the aforementioned method to pre-identify buildings with a high probability of being non-vacant, the number of buildings requiring field surveys can be significantly reduced. This preliminary narrowing down of buildings that are unlikely to be vacant allows municipalities to focus their field survey efforts on a more targeted set of properties. Consequently, this approach is anticipated to substantially alleviate the workload associated with vacant house survey tasks in local governments.

This approach is intrinsically linked to DX, a smart city initiative aimed at enhancing the operational efficiency of municipalities by leveraging their data holdings. It also holds potential for conducting high-frequency, low-cost surveys to better understand the distribution of vacant houses, including those previously overlooked in surveys. Vacant houses that are in good condition and have not been empty for an extended period could potentially re-enter the property market and be renovated (Suzuki and Hino, 2017), thereby improving energy efficiency. Conversely, houses that are initially in good condition but omitted from surveys and left vacant for a long time may deteriorate, becoming challenging to renovate and ultimately contributing to inefficiencies in citywide energy improvement efforts. In addition, both in Japan and internationally, it has been observed that re-listing vacant houses can help stabilise local property markets by increasing housing supply and potentially reducing house prices (Wuyts *et al.*, 2020; Bourne, 2019). The stabilisation of local property markets is expected to encourage the renovation and renewal of housing throughout the region, ultimately leading to enhanced energy efficiency. Therefore, this approach could significantly advance the development of smart energy cities in Japan.

3.7 Expansion of the Method to Other Municipalities

While the above studies are focused on two Japanese municipalities, Kagoshima and Asakura, the author has sought to validate the universality of this methodology by applying it to additional municipalities. Similar approaches were implemented in Maebashi (city), Gunma Prefecture (Tomita *et al.*, 2022; Baba *et al.*, 2021), and Wakayama (city), Wakayama Prefecture (Sayuda *et al.*, 2022). These regions have smaller populations, around the 300,000 mark, compared to Kagoshima, which boasts a population of approximately 600,000. In both cases, it was confirmed that the distribution of vacant houses could be estimated with comparable accuracy to that achieved in Kagoshima.

Moreover, efforts are underway to assess the method's effectiveness in towns and villages with much smaller populations than those previously mentioned. These smaller municipalities often face financial constraints, lower birth rates, and an ageing demographic, which intensifies the challenge of mapping out vacant houses compared to the larger cities noted earlier. In particular, this financial constraint is a stumbling block to efforts to realise smart energy cities in municipalities. It has also been pointed out that smaller municipalities are relatively less aware of energy issues than larger cities, resulting in lower priority being given to initiatives related to smart energy cities (Van and Van, 2013; Humphreys *et al.*, 2018). However, municipalities that fail to address the problem of vacant house adequately and do not provide a good supply of housing can lead to a vicious circle in the future, with an increasingly low birth rate and an ageing population, and financial difficulties that will become more difficult to overcome. As a result, the energy efficiency of the municipality will not only not improve in the future but may even worsen. Therefore, by reducing the burden of surveying vacant houses using the method proposed in this study, I examine whether it is possible to address the above issues by realising a quick and inexpensive survey of the distribution of vacant houses even in small municipalities where it is difficult to conduct on-site surveys of vacant houses.

4. Conclusion

In this chapter, I introduced the work on developing a methodology for estimating the distribution of vacant houses using municipal data, showcasing an actual research endeavour that is undertaken. By leveraging data held by local governments, particularly closed data typically restricted to internal use rather than open data, the estimation of vacant house distribution was achieved with high reliability. The research findings detailed in this chapter are poised to significantly aid in municipal surveys of vacant house distribution. First, the outcomes shared herein are expected to address existing research gaps, notably the challenge of swiftly mapping the spatial distribution of vacant houses across extensive areas. Moreover, this series of studies utilises municipal data, which is consistently updated and maintained by local governments across Japan as part of their routine operations, sparing local governments the task of generating new data for employing the methodologies introduced in this chapter. In addition, the municipal data employed in studies is anticipated to offer a universally applicable method for all Japanese municipalities, given the widespread maintenance of similar data by municipalities nationwide. It is also noteworthy that the collaborative use of such data, originally curated and managed for specific purposes, is producing insightful results that were not initially anticipated.

On the other hand, gathering municipal data from various departments and preparing it for use is not straightforward, as its intended application may be constrained by concerns over personal information protection or judged to be used for purposes other than the intended one. However, the research has underscored that municipal data is highly reliable and frequently updated, being continuously maintained, which renders it invaluable for understanding the dynamics of cities and regions. Going forward, I plan to focus on devising methods to harness municipal data more effectively, exploring currently unconsidered applications, investigating strategies for even more secure management and operation of municipal data, and studying suitable anonymisation techniques for its use. By demonstrating the wide-ranging benefits of utilising municipal data, I anticipate its increased activation in addressing various regional challenges, including the issue of vacant houses, thereby making a significant contribution to regional problem-solving efforts.

The Vacant Houses Special Measures Act specifies that municipal data may be utilised to assess the distribution of vacant houses. However, despite this legal framework, municipal data is not widely employed in local government surveys of vacant houses across Japan. This underuse is attributed not only to institutional barriers, such as the necessity of securing approvals from both the Personal Information Protection Review Board within the municipality and the municipal assembly, but also to psychological barriers faced by municipal staff, assembly members, and residents. These obstacles have hindered the application of municipal data in vacant house surveys. I hope that these efforts will be a pioneering step in breaking through these hurdles and help promote DX in local governments throughout Japan.

Finally, as also outlined in Section 1.3, I anticipate that initiatives will contribute to addressing Japan's urban energy issues in the medium to long term. In 2020, the Japanese government committed to achieving carbon neutrality by 2050. It is expected that there will be a nationwide increase in the installation of renewable energy production facilities within cities. Vacant houses have great potential to be used as spaces for such facilities. In this regard, I believe that vacant houses can simultaneously address both the issues of vacancy and energy, by reimagining them not as negative spaces but as valuable assets for sustainable development.

In addition, as previously mentioned, placing vacant houses on the property market before their condition deteriorates, rather than leaving them unoccupied, and promoting their renovation and renewal can significantly contribute to the realisation of smart energy cities. However, this requires the understanding and cooperation of local residents, industry, and government sectors, including national and local governments and private real estate companies, as vacant houses are primarily assets owned by individuals. In the context of energy issues, there are emerging movements not only in Japan but also in Europe to address regional energy challenges from the bottom up by skilfully forming local communities and encouraging citizen participation, thus fostering sustainable regions (Sokołowski, 2021; Sanada, 2023). By integrating the issue of vacant houses into these efforts, it may be possible to simultaneously advance bottom-up solutions to the problem of vacant housing and initiatives aimed at realising smart energy cities.

Since the nuclear accident following the Great East Japan Earthquake in 2011, the Japanese Government has been undertaking a complete review of its energy strategy, which relies on nuclear power, with the aim of increasing energy self-sufficiency through the implementation of community-based energy networks (Zappa, 2022). The aforementioned initiatives on vacant houses contribute to this activity. In view of the unstable global situation in recent years, including the situation in Ukraine, it can be said that the above initiatives are an urgent necessity for resource-poor Japan. The author also hopes that Japan's future efforts to address the issue of vacant houses and realise smart energy cities will serve as a valuable example for many other countries similarly impacted by the global energy situation. The author is committed to continuing support for this initiative.

Acknowledgements

This study is supported by Policy Research Institute for Land, Infrastructure, Transport and Tourism, Kagoshima City, Asakura City, and the Digital Urban Spatial Information Research and Development Unit, Advanced Research Institute, Tokyo City University. The author expresses deep gratitude to these organisations.

References

Act on Partial Revision of the Vacant Houses Special Measures Act (No. 50 of 2023). In Japanese: 空家等対策の推進に関する特別措置法の一部を改正する法律 [*akiya-tō taisaku no suishin ni kansuru tokubetsu sochi-hō no ichibu o kaisei suru hōritsu*]. Available at: https://www.mlit.go.jp/jutakukentiku/house/jutakukentiku_house_tk3_000138.html (Accessed: 07 January 2025).

Administrative Evaluation Bureau, Ministry of Internal Affairs and Communications (2019) "空き家対策に関する実態調査" [*Report on the Results of a Survey on Measures to Deal With Vacant Houses*]. Available at: https://www.soumu.go.jp/main_content/000595230.pdf (Accessed: 29 April 2024).

Aki, S., Anjali, S.K. and Kenji, I. (2022) "Tokyo Smart Global Megacity: Smart Sustainable Energy Solutions", in T.M. Vinod Kumar (ed) *Smart Global Megacities: Advances in 21st Century Human Settlements*. Singapore: Springer Singapore, pp. 191–218.

Akiyama, Y., Ono, Y., Baba, H., Ueda, A. and Takaoka, H. (2021) "機械学習による空き家分布把握手法の更なる高度化自治体の公共データを活用した空き家の分布把握手法に関する研究(その3)" [*Sophistication of Monitoring Method for Spatial Distribution of Vacant Houses Using Machine Learning – A Study on the Estimation Method of Spatial Distribution of Vacant Houses Using Municipal Public Data (Part 3)*], *Journal of Architectural Planning*, 786, pp. 2136–2146.

Akiyama, Y. and Shibasaki, R. (2015) "ミクロな将来推計人口データを用いた将来の空き家分布推定" [*Estimation of Future Distribution of Vacant Houses Using Micro Future Population Data*], in *CSIS 2015 Research Abstracts on Spatial Information Science*, p. 41. Available at: https://www.csis.u-tokyo.ac.jp/csisdays2015/csisdays2015-ra-pdf/CSISDAYS2015_Abstract.pdf (Accessed: 29 April 2024).

Akiyama, Y., Ueda, A., Ono, Y. Takaoka, H., Kino, Y. and Hisadomi, K. (2018) "鹿児島県鹿児島市における公共データを活用した空き家の分布把握 自治体の公共データを活用した空き家の分布把握手法に関する研究(その1)" [*Monitoring of Spatial Distribution of Vacant Houses Using Municipal Public Data in Kagoshima City, Kagoshima Prefecture – A Study on the Estimation Method of Spatial Distribution of Vacant Houses Using Municipal Public Data (Part 1)*], *Journal of Architectural Planning, AIJ*, 83(744), pp. 275–283.

Akiyama, Y., Ueda, A., Ono, Y. Takaoka, H., Kino, Y. and Hisadomi, K. (2020) "Estimating the Spatial Distribution of Vacant Houses Using Public Municipal Data", in P. Kyriakidis et al. (eds) *Geospatial Technologies for Local and Regional Development*. Cham: Springer, pp. 165–183.

Asami, Y. (2014) 都市の空閑地・空き家を考える [*Considering Vacant Lots and Houses in Cities*]. Tokyo: Progress.

Asano, J. and Inoue, Y. (2022) "地方中核市における空家等対策計画の運用状況と課題に関する研究" [*A Study on the Operational Situation and Issue of Countermeasures Plan for Vacant House in Regional Core City*], *Journal of the city planning institute of Japan*, 57(1), pp. 114–125.

Baba, H., Akiyama, Y. and Yachida, O. (2021) "自治体保有データを活用した空き家の空間分布の将来予測モデル構築 一群馬県前橋市を対象として一" [*Developing Predictive Model for Vacant Housing Distribution Using Municipality-Owned Data: Case Study in Maebashi City*], *Journal of Japan Society of Civil Engineers*, 77(2), pp. 62–71.

Baba, H. and Asami, Y. (2017) "Regional Differences in the Socio-Economic and Built-Environment Factors of Vacant House Ratio as a Key Indicator for Spatial Situation of Shrinking Cities", in *Proceedings of 2017 International Conference of Asian-Pacific Planning Societies, #069*. Available at: https://www.cpij.or.jp/com/iac/upload/file/2017icapps/069.pdf (Accessed: 29 April 2024).

Bernt, M. (2009) "Partnerships for Demolition: The Governance of Urban Renewal in East Germany's Shrinking Cities", *International Journal of Urban and Regional Research*, 33(3), pp. 754–769.

Bourne, J. (2019) "Empty Homes: Mapping the Extent and Value of Low-Use Domestic Property in England and Wales", *Palgrave Communications*, 5(1), pp. 1–14.

Chen, S., Jin, H. and Li, L. (2023) "Analysis and Comparison of House Price Prediction Based on XGboost and LightGBM", *AEMPS*, 46, pp. 55–61.

Chen, T. and Guestrin, C. (2016) "XGBoost: A Scalable Tree Boosting System", in *Proceedings of the 22nd ACM SIGKDD International Conference on Knowledge Discovery and Data Mining*, pp. 785–794.

Couch, C. and Cocks, M. (2013) "Housing Vacancy and the Shrinking City: Trends and Policies in the UK and the City of Liverpool", *Housing Studies*, 28(3), pp. 499–519.

Deng, Q., Shan, M., Zhang, G., Zhang, S., Liu, Y. and Yang, X. (2023) "Effect Evaluation of Large-Scale Energy Saving Renovation of Rural Buildings in Beijing and Implications for Other Cities in the Same Zone", *Sustainability*, 15(6), p. 5580.

Holroyd, C. (2018) *Green Japan: Environmental Technologies, Innovation Policy, and the Pursuit of Green Growth*. Toronto: University of Toronto Press.

Humphreys, E., van der Kerk, A. and Fonseca, C. (2018) "Public Finance for Water Infrastructure Development and Its Practical Challenges for Small Towns", *Water Policy*, 20(S1), pp. 100–111.

Ignacio, A. and Soria, C.D. (2021) "A Study on the Implementation of a Monitoring System to Solar-Powered Houses", *International Journal of Multidisciplinary: Applied Business and Education Research*, 2(10), pp. 902–908.

Imai, A., Sugimoto, S., Sakakibara, W. and Mizuishi, T. (2015) "「空き家問題」の今後と中古住宅の活用可能性" [*Future of the "Vacant House Problem" and the Potential Use of Existing Houses*], *Knowledge Creation and Integration*, 23(8), pp. 20–37. Available at: https://www.nri.com/-/media/Corporate/jp/Files/PDF/knowledge/publication/chitekishisan/2015/08/cs20150804.pdf (Accessed: 29 April 2024).

Jia, N., Wang, J.S. and Li, N. (2012) "Application of Data Mining in Intelligent Power Consumption", in *Proceedings of the International Conference on Automatic Control and Artificial Intelligence. ACAI 2012*, 3–5 March 2012, Xiamen: IET Conference Publications, pp. 538–541.

Joo, H., Lee, S., Kang, S.-J. and Kim, S.-Y. (2022) "Vacant House Characteristics by Use Area and Their Application to Sustainable Community", *Applied Sciences*, 12(21), p. 10696.

Komatsu, Y. and Endo, K. (2000) "Lifetime and Life Cycle Cost Estimation of Japanese Detached House", *Second Japan-Scandinavia Seminar on Building Technology*, 25–26 May 2000, Majvik. Available at: https://ykom.w.waseda.jp/SJS200005.pdf (Accessed: 29 April 2024).

Le, L.T., Nguyen, H., Zhou, J., Dou, J. and Moayedi, H. (2019) "Estimating the Heating Load of Buildings for Smart City Planning Using a Novel Artificial Intelligence Technique PSO-XGBoost", *Applied Science*, 9(13), p. 2714.

Liang, M., Liu, Z., Feng, Z. and Xu, Y. (2020) "Predicting Hard Rock Pillar Stability Using GBDT, XGBoost, and LightGBM Algorithms", *Mathematics*, 8(5), p. 765.

Mashita, M. and Akiyama, Y. (2020) "日本国内における近年の空き家研究の動向" [*The Recent Trend of Housing Vacancy Studies in Japan*], *Journal of Geographical Space*, 13(1), pp. 1–26.

Minami, K. (2021) "Japan's Act Concerning the Promotion of Long-Life Quality Housing", in H.K. Stephen (ed) *Residential Architecture as Infrastructure*. London: Routledge, pp. 201–214.

Mitsunaga, T., Fujiwara, Y. and Okada, S. (2021) "Proposal for Realization and Improvement of DX in Municipal Services Based on Digital Business Models", in *2021 IEEE International Conference on Computing, 2021 ICOCO*, 17–19 November 2021, pp. 158–162.

Mizusawa, K., Tamura, S. and Tanaka, T. (2021) "斜面市街地における空き家の発生要因に関する研究 – 広島県呉市両城地区を対象として" [*A Study on Occurrence Factors of Vacant Houses in the Hilly Residential Area – Case Study of Ryojo Kure City, Hiroshima Prefecture*], *Journal of the City Planning Institute of Japan*, 56(3), pp. 897–904.

Pratheeba, C., Muthuvinayagam, M., Siva Ramkumar, M., Rohith Bhat, C., Maniraj, P. and Kumar, N.S. (2023) "A Review of an Off grid Solar DC System for Rural Houses", in *2023 7th International Conference on Intelligent Computing and Control Systems. 7th ICICCS*, 17–19 May 2023, Madurai, pp. 1837–1843.

Sadayuki, T., Kanayama, Y. and Arimura, T.H. (2020) "The Externality of Vacant Houses: The Case of Toshima Municipality, Tokyo, Japan", *Review of Regional Studies*, 50(2), pp. 260–281.

Sanada, K. (2023) "Smart Cities in Japan and the EU: In Search of Structural Focal Points in Respective Policy Development", *A Journal of the Humanities & Social Sciences*, 27(3), pp. 291–310.

Sasatani, D., Eastin, I.L. and Roos, J.A. (2010) "Emerging Power Builders: Japan's Transitional Housing Industry after the Lost Decade", *CINTRAFOR Working Papers, 57*. Available at: https://digital.lib.washington.edu/researchworks/handle/1773/35448 (Accessed: 29 April 2024).

Sayuda, K., Hong, E., Akiyama, Y., Baba, H., Tokudomi, T. and Akatani, T. (2022) "Accuracy of Vacant Housing Detection Models: An Empirical Evaluation Using Municipal and National Census Datasets", *Transactions in GIS*, 26(7), pp. 3003–3027.

Shinobe H. and Urabe T. (2014) "The Present Conditions and Problems of the Proper Management Regulations of Vacant Houses: As an Example of Local Governments in Eastern Japan", *AIJ Journal of Technology and Design*, 20(45), pp. 723–726.

Silverman, R.M., Yin, L.I. and Patterson, K.L. (2013) "Dawn of the Dead City: An Exploratory Analysis of Vacant Addresses in Buffalo, NY 2008–2010", *Journal of Urban Affairs*, 35(2), pp. 131–152.

So, T. and Arai, Y. (2020) "賃貸住宅入居希望者からの問い合わせがあるかどうか、募集物件情報から予測できるか – 解析手法による予測精度の違いとビジネスへの適用のしやすさに関する考察" [*Predicting Inquiry from Potential Renters Using Property Listing Information – Prediction Accuracy and Applicability to Business*], in *The 34th Annual Conference of the Japanese Society for Artificial Intelligence. 34th JSAI*, 9–12 June 2020, online, 1L3GS1303.

Sokołowski, M.M. (2021) "Models of Energy Communities in Japan (Enekomi): Regulatory Solutions From the European Union (Rescoms and Citencoms)", *European Energy and Environmental Law Review*, 30(4), pp. 149–159.

Statistics Bureau of Japan (2018) "平成30年度住宅・土地統計調査" [*Housing and Land Survey in 2018*].

Sumikura H., Hyodo, S. and Ishigaki A. (2020) "Housing Acquisition and Community Participation by Migrants in Depopulated Areas of Japan: A Case Study in Osaki Kamijima Town, Hiroshima", *IOP Conference Series: Earth and Environmental Science*, 588, pp. 1.11–1.14.

Sun, W.H., Xu, S.C. and Li, L.S. (2011) "Reducing Vacant Houses is More Important Than the Implementation of Green Buildings", *Advanced Materials Research*, 280, pp. 250–254.

Suzuki, H. and Hino, K. (2017) "東京大都市圏の都市部における長期空き家の実態 -川口市空き家実態調査の分析" [*Long-Term Vacant Houses in Urban Areas in the Tokyo Metropolitan Area: Analysis of a Survey of Vacant Houses in Kawaguchi City, Saitama*], *Report of the City Planning Institute of Japan*, 16, pp. 103–107.

Tanikawa, H. and Hashimoto, S. (2009) "Urban Stock Over Time: Spatial Material Stock Analysis Using 4d-GIS", *Building Research & Information*, 37(5–6), pp. 483–502.

Tomita, K., Akiyama, Y., Baba, H. and Yachida, O. (2022) "Estimating the Spatial Distribution of Vacant Houses With Machine Learning Using Municipal Data", in *2022 IEEE International Geoscience and Remote Sensing Symposium, IGARSS 2022*, 17–22 July 2022, Kuala Lumpur: IEEE, pp. 1276–1279.

Ueda, A., Akiyama, Y. and Ono, Y. (2016) "空き家発生・分布メカニズムの解明に関する調査研究(その1)" [*Study on the Mechanism of Occurrence and Distribution of Vacant Houses (Part 1)*], *PRI Review*, 61, pp. 24–35.

Ueda, M. (2023) "Analysis of DX Disincentives in Japan: Qualitative Analysis Through Interviews", in *Proceedings of the 10th Multidisciplinary International Social Networks Conference. MISNC 2023*, 4–6 September 2023. Phuket: Association for Computing Machinery, pp. 124–128.

van der Mescht, J. and van Jaarsveld, M. (2013) "Challenges in Smaller Municipalities-Operations and Maintenance: Municipal Operations", *IMIESA*, 38(2), pp. 17–25.

Wuyts, W., Sedlitzky, R., Morita, M. and Tanikawa, H. (2020) "Understanding and Managing Vacant Houses in Support of a Material Stock-Type Society: The Case of Kitakyushu, Japan", *Sustainability*, 12(13), p. 5363.

Yamada, K., Ii, R., Yamamoto, M., Ueda, H. and Sakai, S. (2023) "Japan's Greenhouse Gas Reduction Scenarios Toward Net Zero by 2050 in the Material Cycles and Waste Management Sector", *Journal of Material Cycles and Waste Management*, 25(4), pp. 1807–1823.

Zappa, M. (2022) "Towards European 'Smart Communities'? EU's Energy Preoccupations and the Lesson of Post-Fukushima Japan", *Istituto Affari Internazionali*, 22/23, pp. 1–13.

7 Legal System and Public Policy of Smart Cities in Energy Transition

Germany – Japan Contexts

Fukuzo Hasegawa

1. Introduction

There are many legal issues in the role of smart cities in the energy transition towards carbon neutrality. Discussion and analysis on this issue are an important theme not only in Japan but also worldwide. This also includes Germany. For example, in 2020 and 2022, the Association of German Jurists (*Deutscher Juristentag*) held discussions on the topic "Sustainable Future Cities and Law" seen from a public legal point of view (Deutscher Juristentag e.V., 2023). This discussion presented enhancement of transportation planning, strengthening of climate protection policy, and addressing residential land policy. On the other hand, issues such as inadequacy of economic perspective and cooperation between local governments and the federal government were also pointed out.

In other parts of the world, the impact of COVID-19 has also changed urban life, as in Germany. It is also very suggestive that the promotion of digital technology has materialised a path to simplifying administrative actions and procedures. It is beneficial to refer to examples and analyses from overseas when solving the legal problems of smart cities in Japan. Therefore, in this chapter, a legal discussion on energy transitions and cities in Germany is offered to propose solutions to the challenges faced by the Japanese smart cities.

2. Current Status of Sustainable Cities in Germany

In Germany, sustainable cities for the future are a prominent topic in literature as the subject of political goals related to the transportation, environment, climate protection, and housing sectors (Burgi, 2022, p. 2726). Going back in time to the discussion on sustainable cities in Germany, as in other developed countries, leads to a report "Our Common Future" issued in 1987. As stated in this report, also known as the Brundtland Report, "[s]ustainable development is development that meets the needs of the present without compromising the ability of future generations to meet their own needs" (World Commission on Environment and Development, 1987, p. 41). One should also note that the

DOI: 10.4324/9781003471448-7

concept of sustainability is not only implemented as a policy at the national level in each European country but also adopted as a guiding principle with an ecological focus.

Referring this to the current event, the major impact on urban life and the legal system in Germany was not only the COVID-19 pandemic and the war in Ukraine but also the flooding of Germany in the summer of 2021. These three significant events dramatically increased the need for changing people's daily lives. These issues were discussed in the article "Eight Theses on How City Life Will Change"[1] published in *Frankfurter Allgemeine Zeitung* in April 2020 (Lembke and Ochs, 2020). The arguments presented in this paper consisted of the following eight items: seductive smart city; back to small parts; sharing as a weakness; caretaker or snitch; city as living room; out to the countryside; nimbly in law; and more space for cyclists. In the context of this growing awareness of these issues, the discussions at *73. Deutschen Juristentages 2020/2022*[2] contributed to the development of a multifaceted analysis and generated several thoughtful comments having importance not only for Germany but also for Japan. The pivot of the discussion at this significant legal forum are summarised in the following section.

3. Discussion at *73. Deutschen Juristentages 2020/2022*

The first issue to be analysed is the need for mobility legislation. A characteristic feature of conventional transportation policy in Germany is the separate legal framework for each means of transportation and type of equipment. Specifically, separate laws regulate the development plans for railroads and roads, and within the field of roads, the institutional regime is further divided between long-distance and peri-urban areas. Based on an awareness of the problem of such fragmentation, Baumgart and Kment (2022) proposed the institutionalisation of "municipal transport planning". While this view emphasises the integration of planning and design in accordance with the needs of municipalities, others expressed the opinion that a federal-wide foundation law should be enacted as soon as possible. For instance, it is the opinion that legislation on mobility at the federal level as a legal framework for an integrated, cross-modal planning and corresponding quality standards (Baumgart and Kment, 2022). Compared to the previous legal situation, which was fragmented by individual modes of transportation (road, rail, etc.), this means applying the idea of codification. At least in urban areas, this concerns using great jurisprudential achievements for the purpose of expanding and adapting the legal framework to sometimes mutually conflicting mobility needs. In other words, the idea is that while public transportation laws and planning designs have traditionally been fragmented by category and type, from the perspective of the permanence and convenience of the transportation services, it would be beneficial from a legal standpoint to create uniform rules. With regard to these conflicting views, it is considered that priority should be given to decision-making at the state and municipal levels, rather than at the federal level, with emphasis on facilitating development planning within cities. Here, it is a worthwhile task to further develop the "communicative common use", which as a legal figure has provided the framework for the pedestrian zones in the inner cities, which today cannot be imagined without, into a "sustainable common use" (Burgi, 2022, p. 2727).

The second issue emphasised at this forum of legal professionals is climate adaptation as a field of activity with an independent legal framework. Particularly considering the 2021 flood in Germany, there was a deeper examination of how cities should respond to climate change. The spectrum of adaptation measures ranges from urban and landscape planning regulations to alleviate climatically induced heating of cities and urban centres, the preservation of fresh air corridors, and even the reduction of the use of open spaces for settlements and infrastructure, to the construction of dikes, the building of backwater flaps, and the reconstruction of sewage systems. As with transportation administration, Germany is characterised by its emphasis on expanding the authority of local governments when it comes to shifting the design of legal systems in response to climate change. A recent specific example is the enactment of a new law in Land Schleswig-Holstein that covers energy transition and climate protection.[3] This new set of articles, revised in December 2021, specifies the goal of climate protection, which is not only German, but also European, and international in scope (Article 1). While this new law is intended to clearly delineate three areas of activity – climate protection, climate adaptation, and energy transition – it has some limitations. This is because of the classification of climate protection as a transversal sector that affects the energy sector, industry, transportation, construction, agriculture, and waste management. These sectoral divisions create tensions with the traditional categorisation of the missions of national organisations. Thus, the challenge will have to be overcome as to how heterogeneous fields of action, corresponding to several separate legal domains, can be normatively linked or, better yet, over-formed in order to achieve climate protection goals (Brüning, 2023, p. 1458).

Finally, as discussed at the 73rd edition of the forum, there is a need for strengthening and developing housing-related land policies. This policy issue has already been the focus of deliberations at the 49th *Deutscher Juristentag* in 1972, with recommendation by the then Federal Construction Minister Hans-Jochen Vogel. Although the ripple effect of this recommendation did not last long, the importance of securing the supply of housing through the development of building sites and the renovation of existing buildings is now increasing. At the 2020–2022 forum, a comprehensive concept for urban development was presented and expectations were expressed for the Building Land Mobilization Act,[4] which was passed by the German parliament (*Bundestag*) in May 2021. In response to these expectations, it has been suggested that it is important to entrench the common interest of housing supply more firmly from a normative perspective and assign a consistent set of instruments to it. On the other hand, there is a view that housing supply will be more difficult rather than easier, given the expected and justified intensification of the conflict between the intolerable pressures caused by problems in large cities and the public interest in climate protection (Burgi, 2022, p. 2728).

4. New Leipzig Charter

As well as the 2020 and 2022 Association of German Jurists conferences, the New Leipzig Charter – *Neue Leipzig Charta* (NLC; BBSR, 2021) is influencing Germany's smart city policy. NLC, which was adopted at the informal meeting of ministers for urban development on 30 November 2020, is intended as a strategic political guiding document for community-oriented urban development in Europe. This charter takes the place of the Leipzig Charter of 2007.

The principle of sustainability that this charter upholds is seen as the maxim of their management and shaping of the urban future (Schlacke and Grotefels, 2022, p. 65; Böhm, 2022, p. 820; Köck, 2020, p. 1), which, according to the three-dimensional concept of sustainability, consists of a balance of ecological, economic, and social concerns with special consideration of the interests of future generations. These concepts themselves had already been discussed from the end of the 20th century to the beginning of the 21st century (Erbguth, 1999, p. 1082; Ronellenfitsch, 2006, p. 385), but remained abstract tools. The NLC is the embodiment of these concepts at the municipal level (Redeker, 2022, p. 1419).

That the NLC is oriented towards strengthening the role of local governments can be read from another document adopted at the same time as the Charter. This document is entitled *Implementation of the New Leipzig Charter in the Framework of a Multi-level Approach: The Continuation of the Urban Agenda for the EU.*[5] This document is aimed to be the basis for further cooperation at local, regional, national, and European levels, in particular through greater use of the European Structural Funds for the years 2021–2027.

In this context, one can place the main points of the NLC. The NLC's subtitle reads: "The Transformative Power of Cities for the Common Good". The common good orientation is to ensure the provision of public services for all population groups, even in shrinking and outlying municipalities. In addition to the sustainability set forth in the old charter (of 2007), climate protection, resilience, pandemics, migration, building culture, and participation are addressed more clearly. These factors are then optimised through digitisation (smart city) and civil society networks. The framework of the new charter consists of three hierarchies at the domain spatial level and an understanding of the urban structure in three dimensions. The heading for such structural design is "[t]he transformative power of European cities". Under this powerful heading, a hierarchy of "urban neighbourhoods, municipal and regional levels including in metropolitan regions" is presented. The three dimensions of the city are "the righteous city", "the green city", and "the productive city", as well as digitalisation is emphasised as a cross-cutting agenda. Furthermore, the series of urban issues raised inevitably relate to housing supply and land policy, as well as to the protection of open space (Kümper, 2021, p. 902).

Nevertheless, some scholars are sceptical about the effectiveness of the NLC, which has organised the issues facing German cities in this way, because of the fragility of the old urban planning system. On the other hand, some argue that the new charter will have a significant impact on German urban planning practice (Battis, 2022, p. 194). The reason for the positive assessment is that the direction of public service provision for the public good is consistent with the essence of municipal self-governance. The social agenda of municipalities has gained weight, not only because of migration, pandemic, housing shortage, and climate protection, including the requirements for a climate-friendly transport policy. In addition to this, there are certain expectations regarding the active support of the German federal government. It is predicted that new urban policies supported by the federal and state governments, including urban development funds, will provide decisive support, which will also generate soft factors such as culture building.

5. Summary of Trends in Germany

These new trends created by the NLC are also reflected in the policies of the Federal Ministry for the Environment, Nature Conservation, Nuclear Safety and Consumer Protection (BMUV). BMUV released a draft "Action Program of Natural Climate Protection"[6] in March 2022. In this action programme, three areas of action that partially overlap in the context of the urban policy can be extracted. The three areas are as follows: green city, protection of soils as carbon reservoirs, and creation of a near-natural water balance ("sponge city"). The first concept, the green city, is an idea that overlaps with that of the NLC as it aims to ensure that cities have sufficient green space by protecting urban ecosystems through enhanced planting and management of natural green spaces (BMUV, 2022, p. 36). Soil protection, the second item listed, focuses on the carbon storage function of urban soils. It is intended to reduce the sealing of the ground by using space-saving architectural methods when constructing buildings (BMUV, 2022, p. 42).

Lastly, the "sponge city" is a very symbolic approach. As the risk of extreme weather events rises and people become more aware of environmental compatibility and resource efficiency, water management in cities is becoming increasingly important. This importance is probably a common challenge not only in Germany but also in other developed countries around the world, including Japan. The widespread sealing of soils by asphalt and concrete has resulted in local water imbalances in many places. As a result, the risk of flooding associated with rainfall is increasing. In addition, the severity of extreme precipitation events is expected to escalate due to climate change (CSC, 2012a, 2012b). To overcome these problems, decentralised stormwater management with measures for storage, infiltration, evaporation, treatment, and utilisation of stormwater has become the focus of attention. The basis for this integrated stormwater management is the Sponge City concept (Reese, 2020, p. 40). In the action program presented by BMUV, the concept of "water-sensitive city" is described (Reese, 2020, p. 38). Most of the rainwater in sponge cities is absorbed by permeable soils or evaporates through cavities and green roofs. In addition, the natural water balance would ensure an adequate groundwater supply. Thereby, sponge cities not only prevent flooding, but also contribute to drought prevention. Thus, it has been shown that the protection of the natural environment in cities is a powerful factor to cope with climate change (Lorenzen, 2023, p. 400).

6. Institutional Background for Smart Cities in Japan

The discussions and analyses in Germany have been considered as an important reference for the development of smart cities in Japan. The following sections examine the prevailing situation and outlook of Japanese smart cities. Fumio Kishida, who was appointed the Prime Minister in October 2021, proposed the Digital Garden City State Initiative[7] as one of his policy ideas. Since he took office, Prime Minister Kishida has made this initiative one of the main features of his administration and has been working to materialise it as a policy.

Nevertheless, this is not the first idea of this type. Going back to the key concepts of each of Japan's previous cabinets, one could find the Garden City State Initiative[8] proposed by the administration of the Prime Minister Masayoshi Ohira. Launched in 1978, it was the first of its kind in Japan. The basic initiative of the Ohira administration was to respond to the contradictions of the economic growth policies that had been in place until the early 1970s, as well as to the increasing anxiety and tension in citizens' lives. It also emphasised the importance of the "garden city" as an automatic restorative device to prevent the trend towards intense urbanisation, i.e. a society in which the advantages of rural villages and the city could be utilised in harmony (Takeno, 2015, p. 128).

In comparison to these 1970s concepts of the pre-digital days, an overview of the Digital Garden City State Initiative of the Kishida administration is based on a set of different assumptions; however, some similarities are also noticed. The social drivers that are the focus of Kishida's initiative are mainly population decline, ageing and depopulation in rural areas, the concentration of people in the Tokyo metropolitan area, and the hollowing out of local industries. In order to solve these issues, the intent of the project is to realise regional revitalisation while maximising the results of conventional regional development efforts. Furthermore, considering the rapid development of new technologies in recent years, digitalisation is the key to solving local social issues and a source of new value (Cabinet Secretariat, 2023). Although more than 40 years have passed, progress of urbanisation remains a major challenge in Japan and harmony between rural and urban areas has been emphasised for many years.

There is another concept that is indispensable in examining smart cities in Japan. It is called Society 5.0. This symbolic keyword emerged in 2016, when in the Basic Plan for Science and Technology formulated by the Cabinet Office, maximising the use of ICT and bringing affluence to people through initiatives that integrate cyberspace and physical space (the real world) were introduced as the fundamental principles. Furthermore, this Basic Plan shared an idea of the "ultra-smart society" as a vision of future life, addressing Society 5.0 as a tool to realise this vision. The vision of ultra-smart society is based on the society in which goods and services are provided to those who need them, when they need them, and in the amount they need them, where the various needs of society are meticulously addressed, where all people can receive high-quality services, and where people can live vibrantly and comfortably, overcoming differences in age, gender, region, language, and other factors (Cabinet Office, 2023a). Society 5.0 is a principle that utilises this new concept of society, and smart cities are supposed to play a crucial role as a showcase for the realisation of Society 5.0 (Takeguchi, 2022, p. 392). Based on this review of the basic policy concepts surrounding Japan's smart cities, the following discussion offers a practical review of actions on creating smart cities in Japan, including these energy-related concepts.

7. Actions on Smart Cities in Japan

Smart city is an important keyword in Japanese urban policy and legal design. This agenda would include those related to energy transformation. Various cities in Japan are implementing this keyword's principles into society by conducting general actions. One example of these actions is policies related to citizen security. I will examine the camera project as one of the methods to provide security.

An indispensable example of smart cities in Japan can be found in the efforts of the city of Kakogawa (Hyogo Prefecture). With a population of approximately 260,000, this Kansai region city has been very progressive in the measures it has taken with this regard. One may find a beginning of Kakogawa's smart city agenda in launching the 2016 Mimamori Camera Project. As city representatives admit, before the installation of security cameras, this city was experiencing a high rate of crimes such as purse-snatching and bicycle theft, and was ranked fourth worst in Hyogo Prefecture for the number of criminal offences per 1,000 people in fiscal year 2016.[9] Under these circumstances at the time, citizens' needs for safety and security were very high. For the child-rearing generation, there was a great deal of interest in ensuring the security of children going to and from school, and for the older adults, it was important to continue to live safely in their familiar neighbourhoods. Based on this background, Kakogawa formulated an administrative plan in 2016, enacted an ordinance on the installation and operation of cameras in September 2017, and concluded an agreement with the competent police department on the proper management and operation of image data in January 2018. As early as 2017, the first year of the project, 900 cameras were installed around elementary schools and along school routes, and the following year the installation of cameras was expanded to parks and major intersections (Yoshida, 2022, p. 100). As a result, by November 2018, the number of criminal offences per 1,000 population improved to below the average in Hyogo Prefecture.

Moreover, Kakogawa has implemented a digital platform called Decidim that was built on free software with the goal of realising the implementation of participatory democracy. Kakogawa was the first city in Japan to introduce this type of platform in 2020; by September 2022, the city had presented 26 topics and had 1,178 registered users (Miyauchi, 2022, p. 21). Most recently, for example, the city has presented an exchange of opinions on an issue of "How to create a lively area around JR Kakogawa Station", aimed at creating a comfortable and pedestrian-friendly community (Kakogawa City, 2023). The advantages of this platform are numerous, but from the perspective of administration, two points are noteworthy: expanding opportunities for participation in policymaking and ensuring transparency in the decision-making process. The traditional Japanese public consultations model consists of merely announcing the opinions that have been submitted, without facilitating any dialogue or exchange of views among the participants. This method also fails to disclose how the opinions of citizens are incorporated or utilised by the local governments. It is true that the number of citizens subscribing to Decidim is not yet large; however, as digitisation progresses, the number of subscribers are expected to increase, and the utilisation of Decidim will expand. In fact, as of April 2024, a total of 18 local governments in Japan have implemented Decidim (Decidim Japan, 2023). As citizens' interest in smart cities grows, one may anticipate that this platform would contribute to the development of effective policies in each government to address specific issues, including decarbonisation.

Specific national policy regarding carbon neutrality and urban administration in Japan is an initiative by the Ministry of Land, Infrastructure, Transport and Tourism (MLIT). With approximately 50% of total carbon dioxide emissions coming from urban activities, there are high expectations for the efforts and contributions of each city governments towards carbon neutrality. With this respect, the MLIT offers a compilation of actual examples of carbon neutrality in Japanese cities categorising a total of seven cities according to three attributes (MLIT, 2023).

The first category is "areal use of energy", and the approach of Sapporo city, Hokkaido, is presented in the 2023 report. The 1972 Winter Olympics host city faced extremely cold winters with heavy snowfall. With a population of close to two million and a large urban area, efficient energy use is essential for decarbonisation. Particularly to effectively distribute thermal energy, a heat pipeline network in the area between Sapporo station and the city centre was constructed. Moreover, Sapporo is promoting the use of district heating by waiving the occupancy fees for the road space where the heat pipes are installed and relaxing the floor-area ratio restrictions for buildings connected to the heat pipes (Sapporo City, 2022). Given the cold climate with high demand for thermal energy, the success of this initiative could have a significant effect on decarbonisation. Since this is a policy targeting a certain area, it would be necessary to create an environment in which the administration can cooperate with different partners. A clear policy could be one of the solutions.

The second category is "coordination of public transport network development and decarbonisation". The policies of cities of Utsunomiya, Saitama, and Sakai fall into this category, but the most notable is probably Utsunomiya (Tochigi Prefecture) with its recently launched tram network. This new policy in Utsunomiya has similar implications to many German cities that keep their tramways (*Straßenbahn*) open for business. In Japan, trams in urban areas have been phased out one after another since the 1960s with the rise of motorisation. At its peak, tram networks covered almost all prefectural capitals, but as of 2022, only 15 prefectures (out of 47 prefectures) had trams in operation.

In this context, Utsunomiya has been planning its transportation in a smart way, with collaboration and aggregation as key elements. Due to limited capacity and frequent delays of bus transportation, as well as its inherent problems of air pollution and low energy efficiency, in 2015 Utsunomiya established Light Rail Co., Ltd. with the city as the largest shareholder (Utsunomiya City, 2023). The purpose of preventing air pollution and ensuring on-time public transportation overlaps with the reasons why German cities value trams. According to interviews conducted by MLIT, the introduction of the trams was also intended to improve the city's image. In the same vein, the city plans to introduce electric buses. Moreover, in conjunction with a series of plans that require a stable supply of electricity, Utsunomiya has been strengthening cooperation with Tokyo Electric Power Company to realise an idea of "local production for local consumption" with respect to electricity (MLIT, 2023).

The third category is "decarbonisation combined with urban renewal". The cities in this category are Odawara, Himeji, and Amagasaki, which are relatively medium-sized cities. Odawara, for example, has long had a serious problem of hollowing out its central city area with vacant houses and stores. Although Odawara flourished as a major post town on the Tokaido road during the Edo period (1603–1868) and has a castle as a cultural asset, the city has failed to make full use of its historical resources. To address these problems, this policy aims to revitalise the central city area by forming a cycle of energy and local economy.

As a base for creating the cycle (starting from decarbonisation), the city of Odawara has established the following three items: (1) maximising the commercial value of locally produced renewable energy; (2) building investment capacity by making energy conservation of electricity and heat energy eligible for grants; and (3) creating an environment that facilitates the participation of local businesses through the expansion of local renewable energy (Odawara City, 2023). The characteristic feature of this project is that it links the traditional issue of revitalising central city areas with the new concept of realising carbon neutrality, leading to effective administrative planning and implementation. This policy, which actively incorporates electric vehicles into car sharing in the downtown area and also focuses on electric shared bicycles, is a progressive one that will improve the city's circulation and economic vitality.

8. Super City and Smart City

What characterises the analysis of smart cities in Japan is a similar framework of super cities. The Act on National Strategic Special Zones[10] was enacted in 2013 to designate zones for focused economic and social structural reform. This act, along with the Act on Special Districts for Structural Reform,[11] was amended in 2020. These amendments aimed to implement cutting-edge technologies in cities to embody the Fourth Industrial Revolution through a problem-solving approach, the cross-cutting use of big data, and regulatory reform and technology implementation using strategic special zones. In this light, super cities, which in content promote digitalisation and pioneering initiatives, are one of the types of smart cities proposed by the Cabinet Office in 2018, although they have different names for them (Terada, 2022, p. 63). The super city concept was behind the initiatives in Toronto where Google cooperates with the Canadian government to understand the movement of people and goods and use big data for urban design, and in Hangzhou, where Alibaba works with the Chinese government to use AI analysis of cameras to combat traffic violations and congestion (Cabinet Office, 2019). Although the background urban environment was forced to undergo major changes due to the COVID-19 pandemic, policy management by the Japanese government and local governments regarding this initiative has continued.

Within this framework, the links between the super cities and the Digital Garden City State Initiative are particularly interesting with regard to the government activities of the last few years. In brief, the Super City Strategic Special Zone can be regarded as one of several policies within the larger framework of the Digital Garden City State Initiative (Cabinet Office, 2023). The policies of the super cities are formulated by the basic local governments (municipalities) on their own initiative and proposed to the Cabinet Office. As of April 2021, a total of 31 municipalities had submitted proposals. Following this, the cabinet's expert committee reviewed these proposals. In April 2022, three local governments were designated as Digital Garden Health Special Zones and two cities were designated as super cities. Of these special designations, the Tsukuba city concept is particularly extensive in its planning. The city aims for resident-oriented urban development with the active participation of its inhabitants and is eager to introduce mobile mobility in earnest and use drones to deliver packages. The city is also considering internet voting and an on-demand mobile advance voting system (Tsukuba City, 2023).

Still, the structure of the Japanese initiative in the discussed area is difficult to understand at first glance, with multiple concepts and initiatives mixed. Fortunately, there are incentives that shed more light on this like the Super City Smart City Forum 2022 that was held in Tokyo in August 2022. This forum featured lectures and exhibits on the innovative efforts of companies and organisations that support the advanced services and data collaboration infrastructure to be implemented, as well as various initiatives for the realisation of smart cities. A noteworthy presentation covered the Super City Concept as a joint project of Osaka Prefecture and the city of Osaka, a host of the World Exposition in 2025. This project is positioned as the third phase of policies for 2026 and beyond and presents plans for two categories: (1) logistics and mobility and (2) medical care and health. Specifically, regarding the transportation category, the plan aims to promote the widespread use of flying cars and the implementation of urban mobility as a service. With respect to the medical sector, the goals are the introduction of artificial intelligence–based medical support and the speeding up of data linkage and services related to care and health (Osaka City, 2022).

9. Discussion and Conclusion

Based on the analysis of examples from Germany and Japan, this section considers prospects for the energy transition and smart cities seen from the perspective of public administration. The administrative branch power (executive) is one of the elements of the separation of powers and the basic concept of a state's governance structure; however, the scope it encompasses is very broad. Today, administrative activities have concrete effects on various aspects of citizens' daily lives. Administrative law as a science explores the principles and precepts common to the massive body of laws, and it organises and analyses the relationship between the people and the public administration. It is tasked not only with investigating the rules governing the activities of administrative power but also with cultivating these rules. The theoretical system of administrative jurisprudence serves as a guideline for the interpretation of articles where there are existing laws, while it serves as a principle of direction in legislation and policymaking where there are no laws (Sakurai and Hashimoto, 2019, p. 2).

In view of these assumptions, I perceive the approach of administrative regulation to the realisation of smart cities and decarbonisation in Japan to be somewhat unsatisfactory. It may go without saying that there are some differences in the concept of legal systems in each country, and there will be differential considerations of the issues depending on whether the country is an island nation or a country within a continent. However, the refinement of the smart city issue in Germany from a jurisprudential point of view is quite useful and advanced. Positive proposals at the 2020 and 2022 legal forums targeting transportation, climate change, and housing development would promote the realisation of smart cities. It is also symbolic that not only are the legal profession and academia making these recommendations but also the national government (Federal Ministry for the Environment, Nature Conservation, Nuclear Safety and Consumer Protection[12]) has been addressing the various issues faced by the German cities. Particularly suggestive is the focus on the soil's ability to store carbon and control water. Japan's super city, garden city, and smart city concepts are also recognised as advanced, but it is desirable to build and systematise a "one step ahead" view in this respect.

In Japan, the promotion of smart cities is supported by urban planning schemes such as land use planning and urban infrastructure development (Murayama and Morimoto, 2022, p. 73). On the other hand, prior studies have indicated that local governments are considering the introduction of new information technologies but are struggling to successfully match them with companies (Katsumata *et al.*, 2021, p. 1420). The survey also found that although the importance of public–private partnerships is recognised in the promotion of smart cities, there are cases where there is not always concrete progress in the continuity of the business model and the setting of service usage fees. In this analysis, the type of financial resources in shortage differs depending on whether the project is private-sector-led or municipality-led, and the typology of financing methods and the establishment of an appropriate management system are necessary (Tsuda *et al.*, 2021, p. 640). One of the perspectives for solving these problems related to the realisation of smart cities is the importance of ensuring that citizens could participate in the formulation of town development planning (Takeguchi, 2022, p. 399).

In order to steadily advance smart cities and decarbonisation over the next decade, it will first be necessary to develop concrete initiatives and regulations by the relevant ministries and agencies. At the same time, it will be important for local governments, such as prefectures and cities, to ensure communication with citizens, not just the intentions of government offices and councils. Effective communication is not necessarily sufficient simply to hear the opinions of citizens at conventional public hearings and town meetings. It will be required for the administration to provide feedback on the opinions and ideas expressed by citizens, and to reconsider and check the project plan. Such necessity and importance can be justified in the following framework based on the theory of administrative jurisprudence. The democratic legitimacy of administrative activities is supported by the intentions of the citizenry. The reasonable and rational needs of citizens are ultimately embodied in administrative actions, and the policies towards smart cities are one such instance. Ensuring transparency in this decision-making process is the theoretical basis for public participation. The purpose of Japanese Administrative Procedure Act[13] is to improve fairness and transparency in administrative operations, thereby promoting the protection of the rights and interests of the public (Article 1(1)). This purpose-setting extends not only to old-fashioned administrative procedures but also to future-oriented administrative management towards smart cities and decarbonisation. Enhancement of feedback from government agencies also has the effect of improving the acceptance of and trust in government activities by the citizens involved in the relevant respective plans and policies. Key words such as public participation and improvement of management functions, which have been repeatedly pointed out over the last decade of heightened analysis and discussion on smart cities, could be strengthened from the perspective of improving transparency to support the initiatives of ministries, agencies, and local governments. By encouraging such reinforcement and cooperation, smart city and carbon neutral policies in Japan will be realised even more seamlessly.

Acknowledgements

This work was supported by JST Moonshot R&D Grant Number JPMJMS2215.

Notes

1 In German: *Acht Thesen, wie sich das Stadtleben verändern wird.*
2 It was a biennial conference on legal and political issues that took place in Hamburg in 2020 and Bonn in 2022 organised by the Association of German Jurists.
3 In German: *Gesetz zur Energiewende und zum Klimaschutz in Schleswig-Holstein.*
4 In German: *Baulandmobilisierungsgesetz.*
5 In German: *Umsetzung der Neuen Leipzig Charta im Rahmeneines Mehrebenenansatzes: Die Fortführung der Urban Agenda für die EU.*
6 In German: *Aktionsprogramm Natürlicher Klimaschutz.*
7 In Japanese: デジタル田園都市国家構想 [*Dejitaru den-en-toshi kokka kousou*].
8 In Japanese: 田園都市国家構想 [*Den-en-toshi kokka kousou*].
9 2016 fiscal year covers the period from 1 April 2016 to 30 March 2017.
10 In Japanese: 国家戦略特別区域法 [*Kokka senryaku tokubetu kuiki hou*].
11 In Japanese: 構造改革特別区域法 [*Kouzou kaikaku tokubetu kuiki hou*].
12 In German: *Bundesministerium für Umwelt, Naturschutz, nukleare Sicherheit und Verbraucherschutz* (BMUV).
13 In Japanese: 行政手続法 [*Gyousei tetsuduki hou*].

References

Battis, U. (2022) "Die Neue Leipzig Charta" [The New Leipzig Charter], *DVBl*, pp. 193–195.

Baumgart, S. and Kment, M. (2022) *Die nachhaltige Stadt der Zukunft – Welche Neuerungen empfehlen sich zu Verkehr, Umweltschutz und Wohnen? Gutachten zum 73. Deutschen Juristentag Hamburg 2020/Bonn 2022* [Sustainable Cities of the Future – What Innovations Are Recommended for Transport, Environmental Protection, and Housing? Report From the 73rd German Lawyers' Conference Hamburg 2020/Bonn 2022]. Munich: Beck Verlag.

Böhm, M. (2022) "Die nachhaltige Stadt der Zukunft" [Sustainable City of the Future], *JZ*, pp. 820–829.

Brüning, C. (2023) "Zum Mehrwert einer Klimaschutzgesetzgebung der Länder am Beispiel Schleswig-Holstein" [On the Added Value of Climate Protection Legislation by the States: Using the Example of Schleswig-Holstein], *NVwZ*, pp. 1458–1463.

Bundesinstitut für Bau-, Stadt- und Raumforschung (BBSR) (2021) "Neue Leipzig-Charta – Die transformative Kraft der Städte für das Gemeinwohl" [New Leipzig Charter – The Transformative Power of Cities for the Common Good], pp. 4–7.

Bundesministerium für Umwelt, Naturschutz, nukleare Sicherheit und Verbraucherschutz (BMUV) [Federal Ministry for the Environment, Nature Conservation, Nuclear Safety and Consumer Protection] (2022) " Erster Entwurf: Aktionsprogramm Natürlicher Klimaschutz" [First Draft: Action Programme for Natural Climate Protection], pp. 36–44.

Burgi, M. (2022) "Die nachhaltige Stadt der Zukunft und das Recht Charta" [The Sustainable City of the Future and the Legal Charter], *NJW*, pp. 2726–2730.

Cabinet Office (2023a) "The Fifth Basic Plan for Science and Technology". Available at: https://www8.cao.go.jp/cstp/kihonkeikaku/index5.html (Accessed: 10 April 2024).

Cabinet Office, Secretariat for Promotion of Regional Development (2019) "About the 'Super City' Concept". Available at: https://www5.cao.go.jp/keizai-shimon/kaigi/special/reform/wg6/190418/pdf/shiryou3-3.pdf (Accessed: 10 April 2024).

Cabinet Office, Secretariat for Promotion of Regional Development (2023) "About Super Cities and Digital Garden Health Special Zones". Available at: https://www.chisou.go.jp/tiiki/kokusentoc/supercity/supercity.pdf (Accessed: 10 April 2024).

Cabinet Secretariat, Secretariat of the Council for Realization of the Digital Garden City State Initiative (2023) "What is the Digital Garden City State Concept?" Available at: https://www.cas.go.jp/jp/seisaku/digitaldenen/about/index.html (Accessed: 10 April 2024).

Climate Service Center (CSC) (2012a) "Machbarkeitsstudie 'Starkregenrisiko 2050' – Abschlussbericht" [Feasibility Study 'Heavy Rain Risk 2050' – Final Report]. Available at: https://www.climate-service-center.de/imperia/md/content/csc/workshopdokumente/extremwetterereignisse/csc_machbarkeitsstudie_abschlussbericht.pdf (Accessed: 10 April 2024).

Climate Service Center (CSC) (2012b) "CSC Report 6, Regionale Klimaprojektionen für Europa und Deutschland" [CSC Report 6, Regional Climate Projections for Europe and Germany]. Available at: https://www.gerics.de/imperia/md/content/csc/projekte/csc-report6.pdf (Accessed: 10 April 2024).

Decidim Japan (2023) "What is MetaDecidim Japan?" Available at: https://meta.diycities.jp/?locale=ja (Accessed: 10 April 2024).

DeutscherJuristentag e.V. (2023) *73. Deutscher Juristentag Bonn 2022* [73rd German Lawyers' Conference Bonn 2022]. Available at: https://djt.de/73-djt/ (Accessed: 10 April 2024).

Erbguth, W. (1999) "Konsequenzen der neueren Rechtsentwicklung im Zeichennachhaltiger Raumentwicklung" [Consequences of Recent Legal Developments in the Context of Sustainable Spatial Development], *DVBl*, pp. 1082–1090.

Kakogawa City (2023) "Kakogawa City Citizen Participatory Consensus Making Platform 'Participatory Process'". Available at: https://kakogawa.diycities.jp/processes (Accessed: 10 April 2024).

Katsumata, W., Kumakura, E. and Shingai, H. (2021) "Survey on Demands for New Technologies Towards Smart Cities to Solve Urban Problems: Questionnaire Survey for Local Authorities Having Use Cases and Demands and Companies Holding Smart City Technologies", *Journal of the City Planning Institute of Japan*, 56(3), pp. 1413–1420.

Köck, W. (2020) "Rechtliche Herausforderungen und Ansätze für eine umweltgerechte und nachhaltige Stadtentwicklung" [Legal Challenges and Approaches for Environmentally Friendly and Sustainable Urban Development], *ZUR*, pp. 1–3.

Kümper, B. (2021) "Wohnraumvorsorge und Freiraumschutz – nachhaltige Flächennutzung durch Baurecht, Bericht über das Symposium des Zentralinstituts für Raumplanung an der Universität Münster am 5. Oktober 2021" [Securing Housing and Protecting Open Spaces: Achieving Sustainable Land Use through Construction Law – Report on the Symposium of the Institute for Spatial Planning at the University of Münster on 5 October 2021], *ZfBR*, 2021, pp. 902–904.

Lembke, J. and Ochs, B. (2020) "Acht Thesen, wie sich das Stadtleben verändern wird" [Eight Theses on How City Life Will Change], *Frankfurter Allgemeine Zeitung*, 26 April 2020. Available at: https://www.faz.net/aktuell/wirtschaft/wohnen/wie-sich-stadtleben-und-wohnen-durch-die-corona-krise-veraendern-16739891.html (Accessed: 10 April 2024).

Lorenzen, J. (2023) "Natürlicher Klimaschutz in der Stadt – Handlungsfelder, Instrumente und Herausforderungen" [Natural Climate Protection in the City – Fields of Action, Instruments and Challenges], *DVBl*, pp. 398–406.

Miyauchi, T. (2022) "Smart City With the Installation of Watching Cameras: Resident Participation Tool 'Decidim' to Engage Young Residents", *Nikkei Glocal*, 447, pp. 20–23.

MLIT, Office of Urban Environmental Policy, Urban Policy Division, Bureau of Urban Affairs (2023) "A Compilation of Case Studies of Efforts Toward Carbon Neutrality in Urban Public Administration". Available at: https://www.mlit.go.jp/toshi/kankyo/toshi_kankyo_fr_000065.html (Accessed: 10 April 2024).

Murayama, A. and Morimoto, A. (2022) "Smart City and Sustainability: From Technology-Oriented to Human and Society-Oriented Sustainable Urban Regeneration", *City Planning Review*, 71(1), pp. 72–75.

Odawara City, Environment Department, Zero Carbon Promotion Division (2023) "Odawara City's Decarbonization Initiatives". Available at: https://www.pref.kanagawa.jp/documents/89493/07odawara.pdf (Accessed: 10 April 2024).

Osaka City (2022) "Osaka Prefecture and Osaka City Super City Concept: Super City Smart City Forum 2022". Available at: https://www.chisou.go.jp/tiiki/kokusentoc/supercity/supercityforum2022/Forum2022_1-01.pdf (Accessed: 10 April 2024).

Redeker, K. (2022) "Die nachhaltige Stadt der Zukunft – Welche Neuregelungenempfehlen sich zu Verkehr, Umweltschutz und Wohnen?" [The Sustainable City of the Future – What New Regulations are Recommended for Transport, Environmental Protection and Housing?], *DVBl*, pp. 1418–1421.

Reese, M. (2020) "Nachhaltiges urbanes Niederschlagsmanagement – Herausforderungen und Rechtsinstrumente" [Sustainable Urban Rainfall Management – Challenges and Legal Instruments], *ZUR*, pp. 40–50.

Ronellenfitsch, M. (2006) "Umwelt und Verkehr unter dem Einfluss des Nachhaltigkeitsprinzips" [Environment and Transport under the Influence of the Sustainability Principle], *NVwZ*, pp. 385–389.

Sakurai, K. and Hashimoto, H. (2019) *Administrative Law*. Tokyo: Yuhikaku.

Sapporo City (2022) "The 2nd Sapporo City Center Energy Plan Promotion Committee Written Report Materials". Available at: https://www.city.sapporo.jp/kikaku/downtown/toshin-energy/documents/r3-02_shiryo01.pdf (Accessed: 10 April 2024).

Schlacke, S. and Grotefels, S. (2022) "Wohnraumvorsorge und Freiraumschutz – nachhaltige Flächennutzung durch Baurecht" [Securing Housing and Protecting Open Spaces: Achieving Sustainable Land Use Through Construction Law], *ZUR*, pp. 65–67.

Takeguchi, K. (2022) "Necessity of Citizen Participation in Building Smart Cities", *Research Bulletin of Naruto University of Education*, 37, pp. 392–402.

Takeno, K. (2015) "Masayoshi Ohira Cabinet as a 'National Garden City Initiative' and Land Planning of Post-War Japan's", *Public Policy and Social Governance*, 3, pp. 125–138.

Terada, M. (2022) "Prospects and Challenges for Smart City/Super City Policies", *Toshi-Mondai Urban Issues*, 113, pp. 63–73.

Tsuda, A., Kawai, T. and Morimoto, A. (2021) "Research on Sustainable Management Organizations for Smart Cities Focusing on Public-Private Partnerships", *Journal of the City Planning Institute of Japan*, 56(3), pp. 635–640.

Tsukuba City (2023) "What is Super City?" Available at: https://www.city.tsukuba.lg.jp/soshikikarasagasu/seisakuinnovationbusmartcitysenryakuka/gyomuannai/2/1013732.html (Accessed: 10 April 2024).

Utsunomiya City (2023) "Haga Utsunomiya LRT". Available at: https://www.city.utsunomiya.tochigi.jp/kurashi/kotsu/lrt/index.html (Accessed: 10 April 2024).
World Commission on Environment and Development (1987) "Report of the World Commission on Environment and Development: Our Common Future". Available at: https://sustainabledevelopment.un.org/content/documents/5987our-common-future.pdf (Accessed: 10 April 2024).
Yoshida, N. (2022) "Using ICT to Solve Local Issues With Citizen Participation: Kakogawa City Smart City Concept", *Governance*, 250, pp. 100–102.

8 Heat Decarbonisation

A Solution for the Future Smart Compact Cities in Japan

Makoto Tajima

1. Introduction

Converting cities into smart cities is often seen as an issue for large cities. Cases of major cities such as Tokyo, Osaka, and Yokohama (Suwa, 2020) are well researched and heavily cited. However, in Japan, it is more relevant socially and environmentally to smaller, local cities.

Decades of ageing and decreasing population presented Japan with a significant challenge, which motivated Japan to shift the city structure to *a compact city* and now to *a smart city*. Due to years of population decline and ageing, many of Japan's municipalities are expected to disappear. In 2014, the Japan Policy Council warned that 896 cities, wards, towns, and villages nationwide are facing the risk of disappearing by 2040. Of these, 523 municipalities will have a population of less than 10,000, making them even more likely to disappear (Masuda, 2014). The decline of municipalities is ongoing. The Japanese population is shrinking by a million people annually. The National Institute of Population and Social Security Research predicts that Japan's population will dip below 100 million by 2056 and halve to approximately 63 million by 2100 (National Institute of Population and Social Security Research, 2024). That affects municipality management. Yubari City, for instance, went bankrupt in 2007 and is now in the process of becoming a restructuring organisation. The government has advocated the *compact city* concept for decades to tackle this issue, but progress has been limited.

Large cities present another issue in the present decarbonisation age. Large cities are major energy consumers and emit significant amounts of greenhouse gases. On the other hand, local cities play a central role in local politics, economics, and culture, and local areas serve as a source of renewable energy for large cities, leading to large cities' dependence on local cities' capacity as energy suppliers.

Revitalising rural areas will also help reduce the overpopulation in large cities and alleviate the problems associated with this concentration. The economic burden on the younger generation is also increasing due to the ageing population. A compact city is an effective way to reduce these social

DOI: 10.4324/9781003471448-8

costs (Nishiura, 2019). Attracting the population into regional compact cities reduces the social costs and rationalises public services. Making the local, compact city into smart city further enhances its function.

2. Compact City

The term *compact city* was first introduced to Japan in 1974, and its primary research was done in the 1990s. The national and local governments have practised it since the 2000s as an urban image against the spreading trend (Kaido, 2001). In December 2006, the Council for Social Infrastructure Development publicly announced the direction of compact cities in Japan for the first time.

Compact cities are compatible with decentralised variable renewable energy. Potential renewable energy sources are primarily concentrated in rural areas due to land constraints (OECD, 2012). While 240 Japanese municipalities have developed master plans for compact cities in Japan, only a handful describe compact cities solely regarding local economy and energy efficiency. Most cities associate them with livability (Koshikawa *et al.*, 2017).

Copenhagen, the capital city of Denmark, serves as an exemplary compact city in Europe. It is transitioning into a smart city through the renowned Finger Plan. This urban development strategy, established in 1948, has effectively created a well-functioning metropolitan area while curbing urban sprawl. Notably, Copenhagen's widely recognised district heating network, covering over 98% of domestic users, is rooted in the principles of the Finger Plan (Figure 8.1) and extends to 63% of the nation.

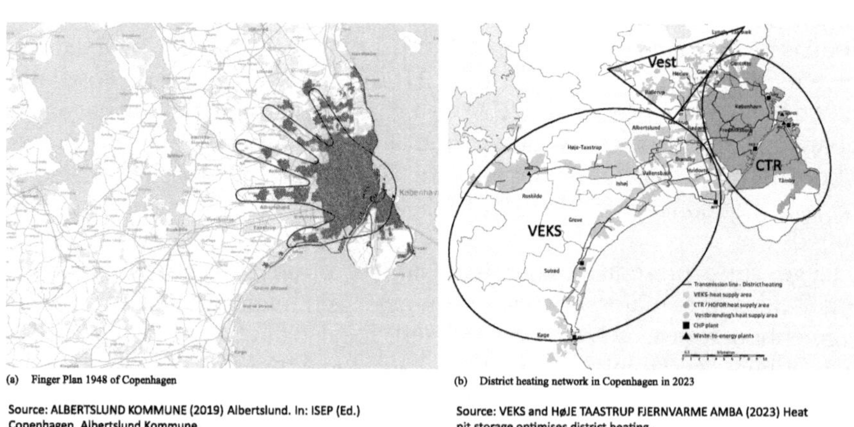

Figure 8.1 Finger plan and today's district heating network in Copenhagen: (a) Finger Plan 1948 of Copenhagen; (b) District heating network in Copenhagen in 2023

Source: (a) Albertslund Kommune (2019); (b) VEKS and Høje Taastrup fjernvarme amba (2023)

3. Japan's Green Transformation and Regional Decarbonisation

The Japanese government aims to achieve net-zero emissions by 2050. It has set a 2030 greenhouse gas reduction target of 46% from its 2013 levels, which requires continued strenuous efforts to cut by 50% from the selected baseline (The Government of Japan, 2023; Ministry of Foreign Affairs of Japan, 2021). This involves a mindset shift towards proactive climate measures that transform the economy and society with proactive policies and regulatory tools (Sokołowski, 2022).

In this context, Japan's key driver for the green transition was the introduction of the feed-in tariff (FIT) scheme in 2012. FIT significantly affected the proliferation of renewable energy in Japan. It was amended in 2017, and the second amendment was enforced in April 2022 (Figure 8.2). Other policy instruments include the latest Sixth National Basic Energy plan, which increased the renewable energy target, and the revised Global Warming Measures Law, which mandates the local government to formulate a renewable energy plan and its target (Figure 8.2). The latter measures will make the local-level intervention more effective.

Following these policy changes, the Ministry of the Environment took a bold move to launch a large-scale grant programme to accelerate the decarbonisation of Japan through "Regional Decarbonisation Transition and Renewable Energy Promotion Grants" (Figure 8.3).

Local governments and other entities ambitious in their decarbonisation efforts will be supported through the programme. The Ministry will select 100 leading municipalities and grant 5 billion JPY or 35 million USD per

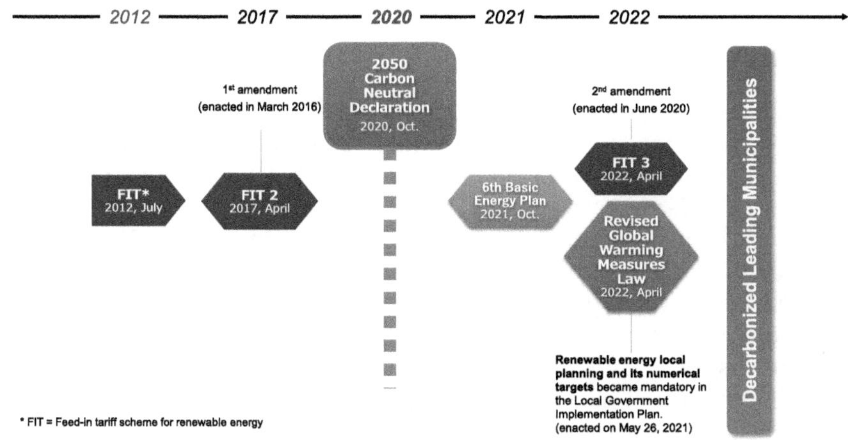

Figure 8.2. Milestones and major drivers in the Japanese green transformation

Source: Produced from the information from the Ministry of the Environment, the Ministry of Agriculture, Forestry and Fisheries, and the Ministry of Economy, Trade and Industry.

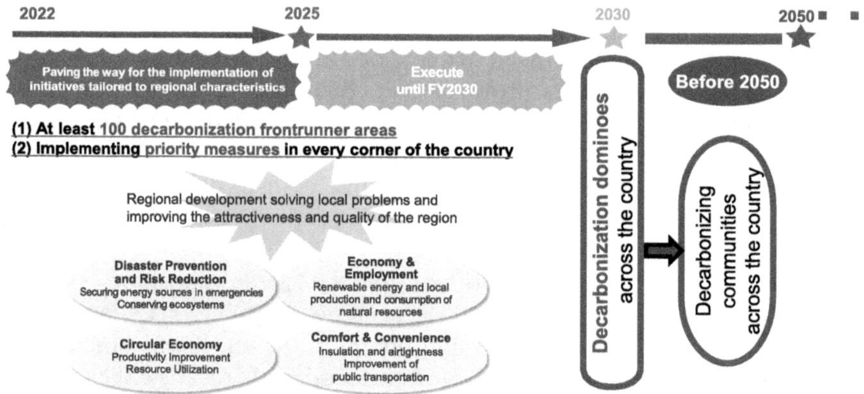

Figure 8.3 National government's initiative to promote regional decarbonisation transition and renewable energy

Source: Ministry of the Environment (2021)

municipality for five years. Those municipalities are expected to become models that others can follow. This programme aims not only to accelerate decarbonisation but also to promote (1) disaster prevention and risk reduction, (2) circular economy, (3) local economy and employment, and (4) the comfort and convenience of the local people. Numerous technical advances and innovations are required to achieve the programme's goal. District heating is highlighted in the following section as an example of a sustainable smart-compact city that can contribute to this goal.

4. District Heating

Heat is a major but often neglected energy use in the decarbonisation debate. However, half of the energy we consume is in the form of heat. About 52% of the total energy consumed in Japanese households is heat, compared to 48% for electricity. Kerosene consumption is exceptionally high in the northern regions: 59% in Hokkaido and 44% in Tohoku (Figure 8.4). These figures do not account for electricity used as heat, which means the actual energy consumption is even higher. In Japanese households, electric air heat pumps or air conditioners are commonly used for room heating during winter. In addition, electricity is also used to supply hot water in some households.

The importance of heat is also evident from a national total consumption. Heat is the largest contributor to energy consumption, accounting for 54.1%, transport fuels for 23.2%, and electricity for 22.7% (Table 8.1). This tendency applies to other countries worldwide. Heating and cooling needs account for the largest global share at 49% of final energy consumption (IRENA *et al.*, 2020).

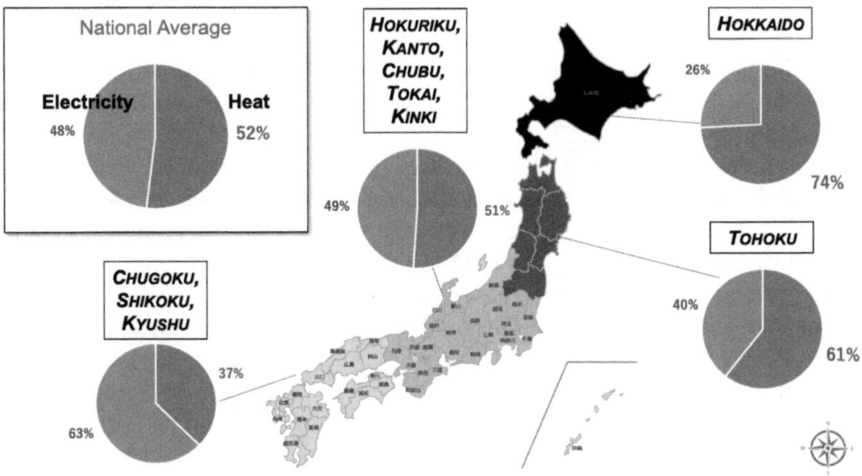

Figure 8.4 The energy source of Japanese household consumption[1]
Source: Produced from Ministry of the Environment (2019)

Table 8.1 National energy consumption by energy carrier

Energy carrier	*Petajoules (PJ)*	*Ratio to the total national energy consumption (%)*
Heat	7,760	54.1
Transport Fuels	3,250	23.2
Electricity	3,330	22.7

Source: Fourth Generation District Heating Forum and Institute for Sustainable Energy Policies (2020).

District heating is well developed in Denmark. Denmark's district heating history began with the construction of a waste incineration plant in 1903. This plant made waste processing more sustainable while providing electricity and steam heat to a new hospital. In the 1920s and 1930s, collective district heating systems were developed based on waste heat from local electricity production. After the oil crisis in the 1970s, district heating from combined heat and power (CHP) expanded rapidly (Sokołowski, 2020). During the energy crisis in 1973–1974, energy consumption per capita had risen to a very high level. The oil crisis created an urgent need to save energy to reduce Denmark's dependency on imported fuels and consumers' heating expenses. As a result, Denmark decided to expand the fuel-efficient CHP system from large to medium and small-sized cities. In the late 1970s, around 30% of all homes in the larger Danish cities were heated by district heating systems (State of Green, 2020). Now, district heating is distributed in more than 400 places throughout the country, covering over 66% of households nationwide, and 93% in Metropolitan Copenhagen are connected to the district heating system. Approximately 120 petajoules (PJ) of distributed heat are used annually for space heating and hot water supply.

Denmark's widespread use of district heating and CHP significantly increased energy efficiency and decreased carbon emissions over the past several decades. Danish district heating, which has 100 years of history, constantly advances. Now, it is getting into the fourth generation to achieve a low-temperature hot water supply to minimise heat loss and enable more use of low-quality heat, such as heat from waste incineration, data centres, factories, and alike, which is dumped otherwise (Figure 8.5). The fourth generation district heating focuses on energy efficiency, flexibility, and integrating all available renewable and surplus energy sources.

Renewable sources gradually replace fossil fuel-based heat sources, and more waste heat sources are integrated into the system as the generation advances. At the same time, a large-scale seasonal thermal storage system became famous as a part of a large-scale solar thermal plant to shift the peak season of solar energy to use surplus heat generated during summer in winter (Sifnaios *et al.*, 2023). This system also stores surplus electricity generated in Denmark, primarily by wind turbines. During periods of low or even negative market prices for electricity, surplus electricity is converted into heat and stored in seasonal thermal storage for use in the district heating network (Tian *et al.*, 2019). This sector coupling, integrating different energy sectors such as electricity, heating, and gas, further contributes to achieving a climate-neutral future via a smart thermal grid (Johansen and Werner, 2022).

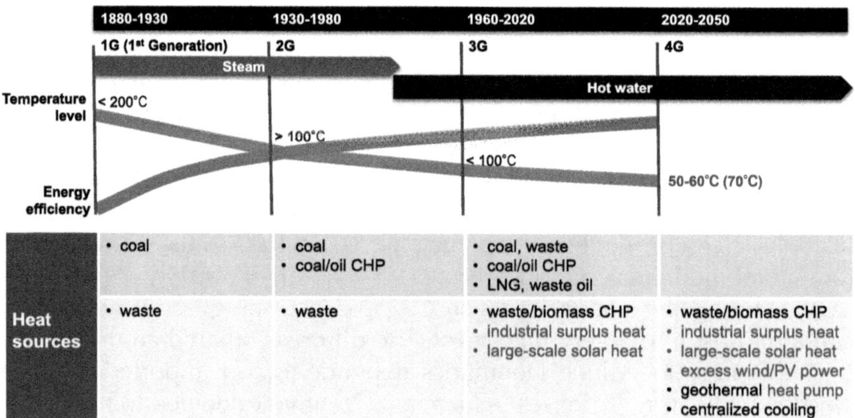

Figure 8.5 Historical development of Danish district heating networks

Source: Produced from the information from Aalborg University and Danfoss District Energy.

A significant benefit of low-temperature district heating systems is their ability to integrate, store, and reuse low-quality heat that would otherwise be wasted, thereby reducing the adverse impact on the environment (Danfoss, 2023). In Japan, 4,814 PJ of excess or surplus heat are wasted annually, nearly twice the amount in Germany and 35 times more than in Denmark (Figure 8.6). Japan also releases approximately 54%, or 9,930 PJ, of its total annual primary energy supply of 18,000 PJ into the atmosphere and ocean as waste heat. The annual cost of fossil fuel imports, which amounts to 24 trillion JPY, could be significantly reduced if Japan efficiently harnesses and utilises waste heat.

The Danish district heating system is so advanced that the European Union (EU) adopted it as a core model for its heat policy, which is essential to achieving the 2050 decarbonisation target (Connolly *et al.*, 2013). District heating is expected to contribute significantly to Europe's reduced fossil fuel–based heat supply and associated decarbonisation.

Danish district heating is also characterised by its unique ownership model (Johansen and Werner, 2022). Most of the Danish district heating is operated by community-owned companies. Communal heat solution is socially and economically robust, sharing risks and rewards among community members and it is well applicable to other countries (Chittum and Østergaard, 2014). *Community Energy* or *Community Power* is widespread in several countries in Western and Northern Europe (IRENA Coalition for Action, 2020). The first renewable energy project in which local stakeholders worked together

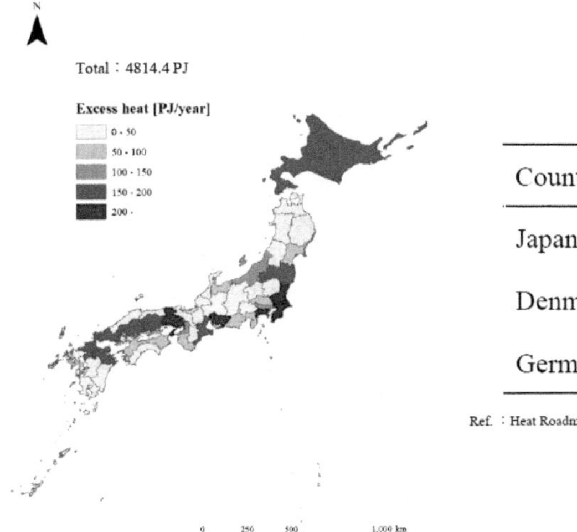

Country	Excess heat [PJ]
Japan	4814
Denmark	139
Germany	2707

Ref. : Heat Roadmap Europe, Background Report 7 – Potential for Excess Heat, 2015.

Figure 8.6 Comparison of the amount of excess heat between Japan, Denmark, and Germany

Source: Nakata and Ozaki (2019).

originated in Danish cooperatives in the 1980s, spread to Germany and other European countries from the 1990s to the 2000s, and, in the 2010s, became active in Canada, Japan, and Australia. In the 2010s, they became active in Canada, Japan, and Australia, and in recent years, they have been spreading to South America and Africa. To promote these trends, the World Wind Energy Association consulted with practitioners and researchers around the world and published the following *Three Principles of Community Power* in 2011 (Iida *et al.*, 2014):

1. Local stakeholders own most or all the project.
2. Project decisions are made by community-based organisations.
3. Most or all the social and economic benefits are distributed to the community.

5. Status of District Heating in Japan

In Japan, district heating was first introduced in Senri New Town in 1970 and later in Shinjuku New City Centre and Sapporo. These initiatives aimed to prevent air pollution before the Osaka Expo and the Sapporo Olympics. Since then, district heating systems have expanded nationwide, with 74 companies operating in 132 districts by the end of FY 2018.

However, the heat supply business constitutes only 0.01% of the total area (compared to 100% for electricity and 5.7% for gas). In terms of customers and annual sales, there are approximately 36,000 district heating customers (equivalent to around 150 billion JPY) compared to 29 million gas customers (amounting to 3.7 trillion JPY) (Table 8.2). This stark difference contrasts with countries like Denmark, where district heating serves over 60% of its customers.

Three key factors contribute to this disparity. First, Japan's historical energy policy has focused on industries such as coal, oil, gas, electricity, and nuclear power, leaving heating and cooling for homes and commercial buildings to the competitive landscape of various energy industries. Second, Japanese buildings, often poorly insulated and airtight, favour *spot heating* (heating or cooling specific rooms) over whole-house solutions due to cost considerations. Third, the high initial installation costs of heat pipes in Japan, several to dozens of times higher than in Europe, pose a challenge. Despite these obstacles, district heating remains an area of interest and potential growth in Japan.

The linear heat density is a crucial economic viability indicator for district heating networks. It is defined as the annual heat demand per meter of grid length. The higher the density, the more heat can be delivered and sold over a grid unit, reducing losses and investment costs (Dochev *et al.*, 2018). In Japan, district heating is mainly introduced for business users in urban centres with high linear heat density. As a result, the annual amount of heat sold through district heating in Japan is one-tenth the size of Germany and one-fifth the size of Denmark (Table 8.3). Tokyo accounts for more than 60% of the total heat sold in Japan (IEA, 2013).

Table 8.2 Comparison of Japanese energy markets: electricity, city gas, LPG, and district heating

	Electricity	*City gas*	*LPG*	*District heating*
Supply area share	100%	6%	94%	0.01%
No. of consumers	84 million	29 million	24 million	36 thousand
Market scale	18 trillion JPY	5 trillion JPY	n.a.	140 billion JPY
No. of employees	130,000	32,400	n.a.	2,500
Heat-pipe distance	-	259,000 km	-	646 km
Time of liberalisation	FY2016	FY2017	Liberated	FY2016
Notes	Over 600 retail electricity suppliers (10 former general electricity utilities)	Approx. 200 companies	Approx. 21,000 companies	76 companies, 134 regions (2017)

Source: Compiled from METI data and other sources.

Table 8.3 Comparison of district heating between Japan and European countries[2]

Index	*Tokyo*	*Japan*	*UK*	*Germany*	*Denmark*
Population	13,960,000	125,570,000	66,650,000	83,020,000	5,810,000
Annual heat sales (GWh/yr)	3,760	6,013	14,173**	75,119	34,538
Annual Sales (billion JPY)		139.1	57.2*	704.6*	312*
Total length of heat pipes (km)	293	646	308	21,610	30,800
Total number of district heating	87	137	2,000	1,454	400
Linear heat density (MWh/m/yr)	11.9	9.3	46.0	3.5	1.1
Renewable energy ratio	5%	15%	2%	12%	59%

Source: Compiled from EUROHEAT&POWER Country by Country 2019, Handbook of Heat Supply Businesses (Japan Heat Supply Association, 2021), results reports, etc. (Institute for Sustainable Energy Policies, 2021)

Denmark has more than 100 times the total length of heat pipes compared to Tokyo, despite having a population less than half that of Tokyo. Its annual sales volume is about ten times greater, resulting in a linear heat density of about one-tenth. Denmark has achieved a sustainable district heating business model, covering more than 60% of the population, including residences.

Germany's linear heat density is between that of Tokyo and Japan and that of Denmark. The United Kingdom is in the same situation as Tokyo and Japan in terms of introducing district heating (Furubayashi and Nakata, 2021).

Denmark has already achieved a sustainable district heating business at a very low linear heat density of 1.1 MWh/m/year, while Japan needs a much higher density of 9.3 (Table 8.3). Low linear heat density is the biggest bottleneck in developing expensive district heating infrastructure in rural areas or local cities. To make the district heating business viable, you need to have relatively high population density or linear heat density. Compact city presents a potential solution to this problem. Developing basic infrastructure in the local smart-compact city could be a rational governmental policy, given Japan's needs to shift from a diffuse to a concentrated city structure.

A potential model for a smart-compact city for local municipalities, the Ogata model, is discussed in the following section.

6. Ogata Model

Ogata Village is potentially a good showcase for a smart-compact city utilising locally available agricultural waste for district heating. Hachirogata Lagoon, located in northwestern Akita Prefecture, was once Japan's second-largest lake after Lake Biwa and was converted to land through the National Hachirogata Lagoon Reclamation Project, which started in the late 1950s (Figure 8.7). The place was named Ogata Village, recruiting immigrant farmers nationwide. Ogata Village can be considered *a compact city* in the present term. In contrast to typical rural areas in Japan, this village was meticulously planned. A central public facility zone, known as the *centre belt,* was established in the city's heart. Village offices, schools, stores, and other public amenities were strategically concentrated near the residential areas.

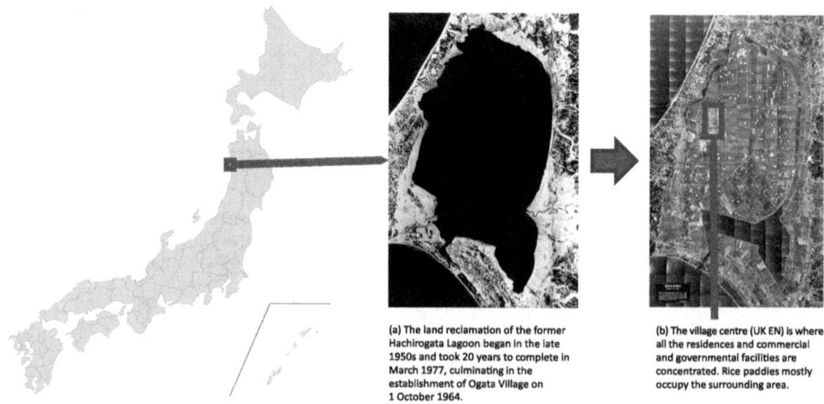

Figure 8.7 Ogata Village, formally Ogata Lagoon, in Akita Prefecture in northern Japan

Source: Produced from the information from the Ogata Village Office (2023)

The basic statistics of Ogata Village are summarised in Table 8.4 and Figure 8.8. The village has a population of approximately 3,000, with 80% of the residents working as full-time rice farmers. The total farmland area is around 9,000 hectares, with an average of about 18 hectares per household. Annual rice production amounts to approximately 61,000 tons. The annual rice production value in the village is approximately 11 billion JPY, ranking sixth among all municipalities in Japan or third in Akita Prefecture. It represents an 11% share of the national market. The village requires approximately 12,000 tons of rice husk processing annually.

Public facilities in the village will be heated using specially developed rice husk biomass boilers installed at the Country Elevator Corporation land (two 350kW boilers introduced in FY2023, four boilers in total by FY2026), which is the site where rice husks are discharged. The boilers are fuelled by 1,500–2,000 t/year rice husk per unit. The heat is transported through

Table 8.4 Profile of Ogata Village

Population	2,992 persons
Households	1,131 households
Primary occupation	80% farming
Farming area	18 ha/household
Primary product	Paddy rice
Annual rice yield	61,000 tons/year
Annual rice husk produced	12,000 tons/year

Source: Ogata Village Office (2023)

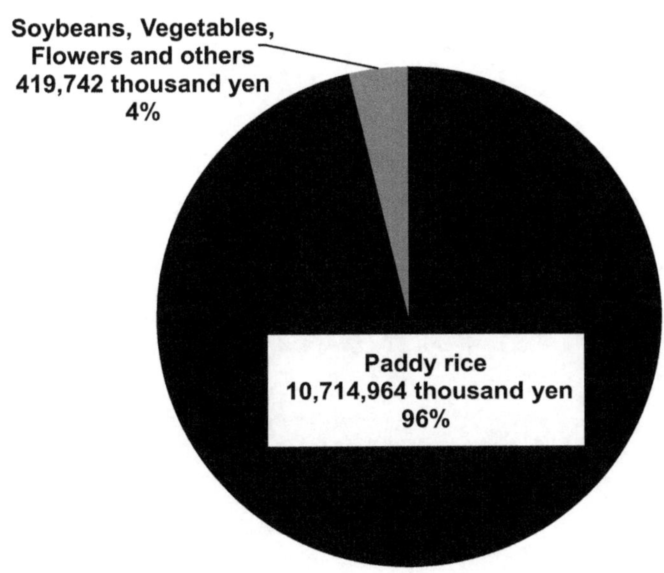

Figure 8.8 Agricultural production in Ogata Village in 2015

Source: Ogata Village Office (2023)

a 3.5-km one-way or 7 km both-way pipeline (Figure 8.9). The project has received technical cooperation from the Danish counterparts in planning, procurement, equipment supply, and installation. It is Japan's first full-fledged replication of the Danish community-based district heating model. Installed

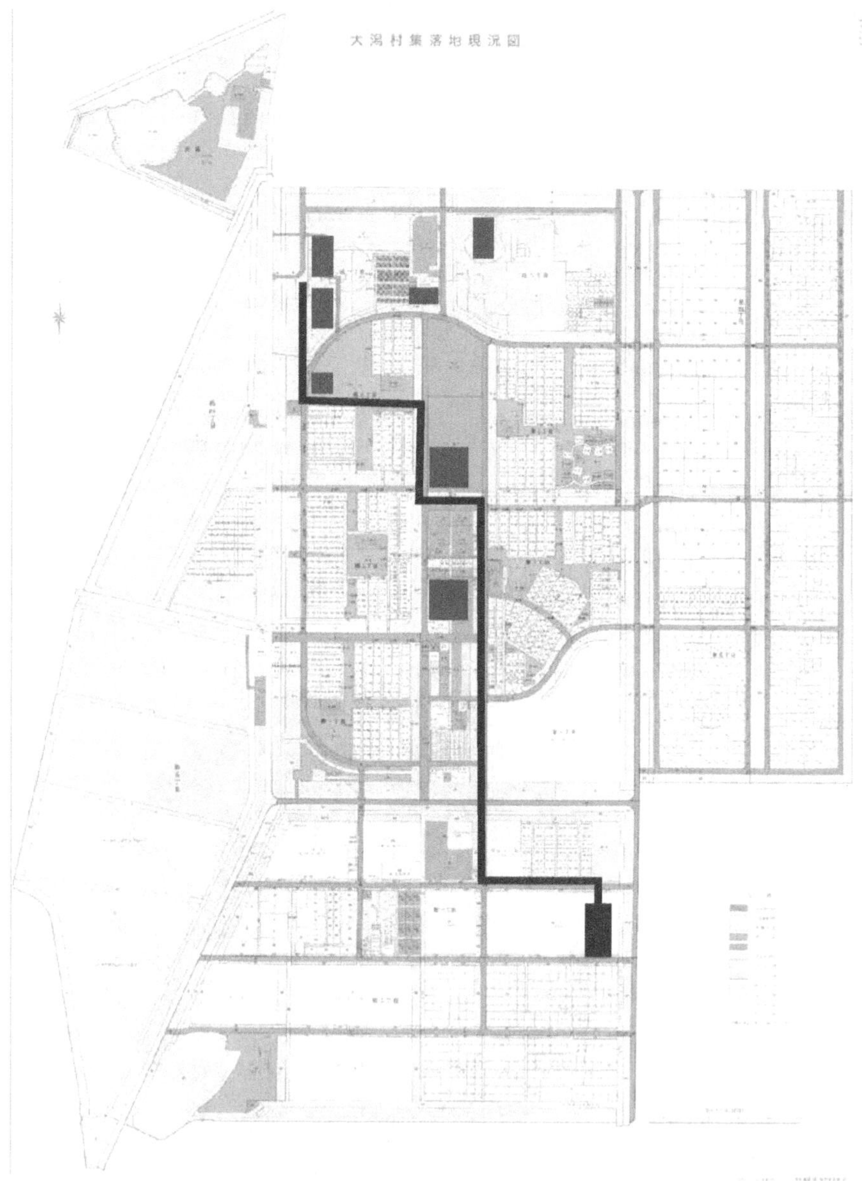

Figure 8.9a The initial district heating system in Ogata Village, Akita Prefecture, Japan (overview)[3]

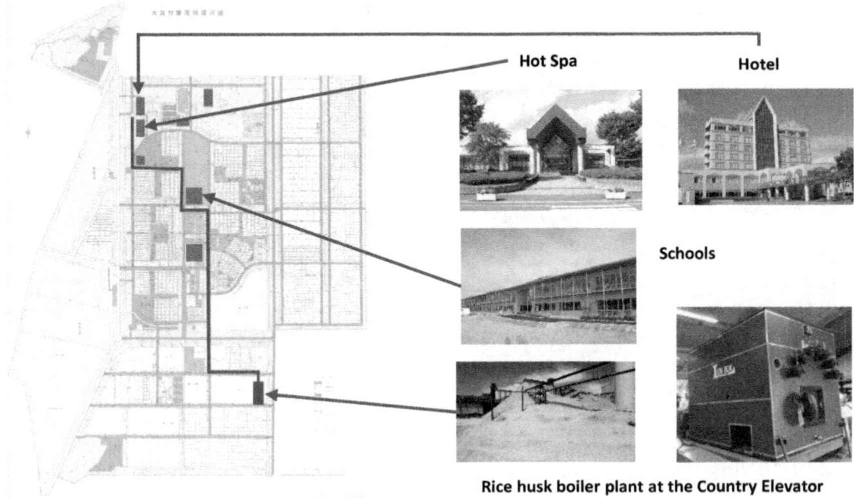

Figure 8.9b The initial district heating system in Ogata Village, Akita Prefecture, Japan[3]

equipment is advanced and intelligent enough to communicate with each other, and almost all operations can be conducted unattended. An operator can check the status remotely with his smartphone or tablet. Warning signals can be received from anywhere worldwide, including Danish companies that monitor boiler and heat pipe network operations.

Heat pipes impose a major installation cost; minimising operation and maintenance costs is the key to a sustainable operation. Danish heat pipes are so advanced that to reduce heat loss during water transportation, you can detect water leakage through the built-in detection mechanism, which can be monitored anywhere in the world.

Denmark's extensive experience developing wheat straw-based biomass boilers and plants led the project to select a Danish boiler, given the similar chemical properties of wheat straw and rice husk. However, a significant challenge arose due to the unique properties of rice husk. This material contains a substantial amount of silica, which crystallises when exposed to high temperatures. In June 2003, based on the International Agency for Research on Cancer recommendation, the European Union Scientific Committee recognised that crystallised silica, mostly cristobalite, has a relative cancer risk to individuals with silicosis, especially those working in the mining, cement, ceramics, glass, and mineral wool, foundry, and precast concrete industry (European Network on Silica, 2020). European Union member countries have established occupational exposure limits ranging from 0.05 to 0.3 mg/m^3. Overcoming this challenge required eight years of collaboration with our Danish counterparts. Together, the project achieved optimal incineration conditions and boiler design to suppress the production of crystallised silica.

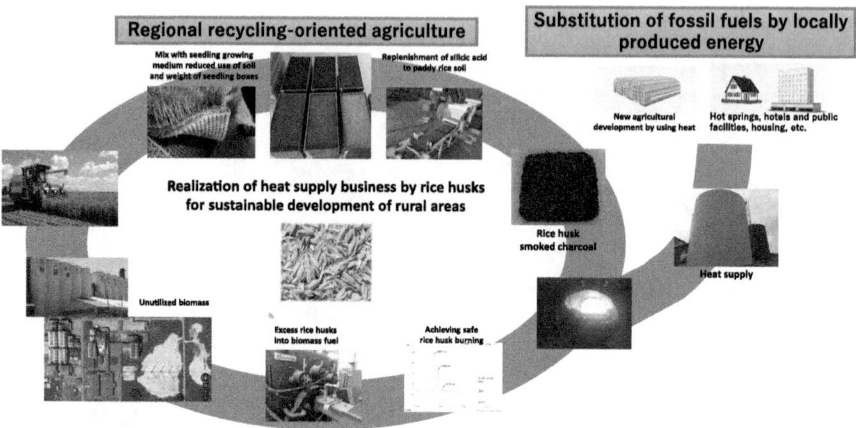

Figure 8.10 Sustainable Ogata Village – recycling-oriented agriculture and locally produced energy.

Though only selected public facilities receive heat initially, the primary system is designed to accommodate future expansion to cover other facilities, including private homes. The Ogata project also envisages a circular economy and recycling (Figure 8.10). Smoked charcoal produced from the rice husk boilers is returned to the farmland by mixing it with seedling bedding soil. Rice farmers in the village buy nursery soils from nearby hills. However, these soils are expected to be depleted soon. To address this, mixing rice husk charcoal into the soil serves two purposes: it conserves the soil and replenishes nutrients, including essential silicic acid.

The Ogata Village project also learned from Denmark in the social aspect and established a local energy company, Ogata Renewable Energy Services, Inc. While there are so-called new electric power companies in Japan involved in the electricity retail business, this is probably the first energy company that handles both electricity and heat, which is essential to achieve efficient energy management through sector coupling.

7. Constraints

We need to address additional constraints when transferring this model to different areas. Some of these constraints necessitate national-level intervention and policy direction.

7.1 Economic and Policy Constraints

Despite the subsidy, the Ogata project faces challenges in making a solid business case. The Japanese government's inadequate guidance on carbon pricing policy has inadvertently created a reverse economic incentive. Unlike

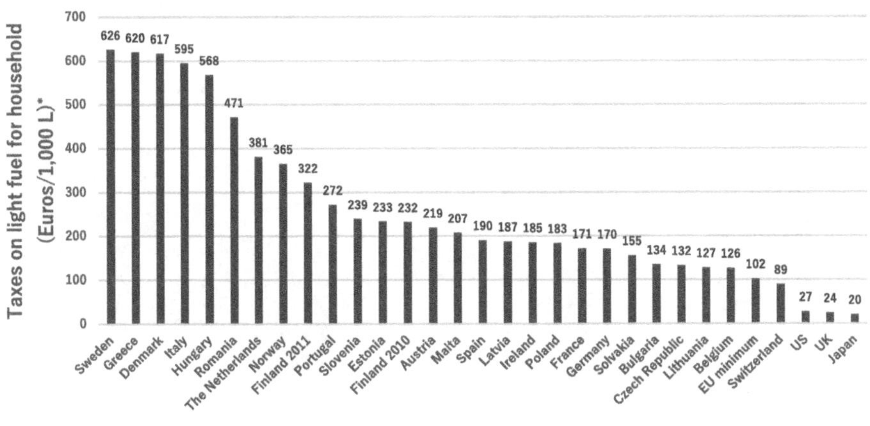

Figure 8.11 Taxes on light fuel oil for households, EUR/1,000 l.

Source: Finnish Energy Industries (2010) "Energy Taxation in Europe, Japan and the United States". Available at: https://dokumen.tips/documents/energy-taxation-in-europe-japanand-the-united-states-of-the-energy-taxation-survey.html?page=1 (Accessed: 30 May 2024).

European countries, Japan's fossil fuel tax is virtually non-existent, especially compared to Denmark's robust system (Figure 8.11). For instance, Japan's domestic fossil fuel tax on light fuel oil for domestic use is a mere fraction of Denmark's rate – 20 EUR per 1,000 litres versus Denmark's 617 EUR. This disparity is further exacerbated by a massive subsidy that the Japanese government provides to maintain stable, low fossil fuel prices. The situation worsened during the 2023 energy crisis triggered by the Russian invasion of Ukraine.

Consequently, the Japanese renewable energy sector faces a significant disadvantage as it competes with artificially cheap fossil fuels. Japan urgently needs a fair carbon pricing mechanism to promote renewable energy and decarbonisation to level the playing field. In addition, the government must address the substantial initial investment costs associated with expensive heat pipe infrastructure or distribution networks, similar to what has been done in European cases.

7.2 Technical and Engineering Constraints

Another constraint is a weak history and limited market of district heating (Table 8.2), which translates into (1) a weak industrial base, (2) a poor accumulation of knowledge, and (3) a lack of skilled personnel and companies. These challenges affect project implementation. Costs and time are increased due to the need for imported equipment, expatriate dispatch, and additional training. Considering over 60 to 100 years of district heating development history in Denmark, it is natural, but there are many technical gaps between Denmark and Japan. Japan needs to address these issues to localise the

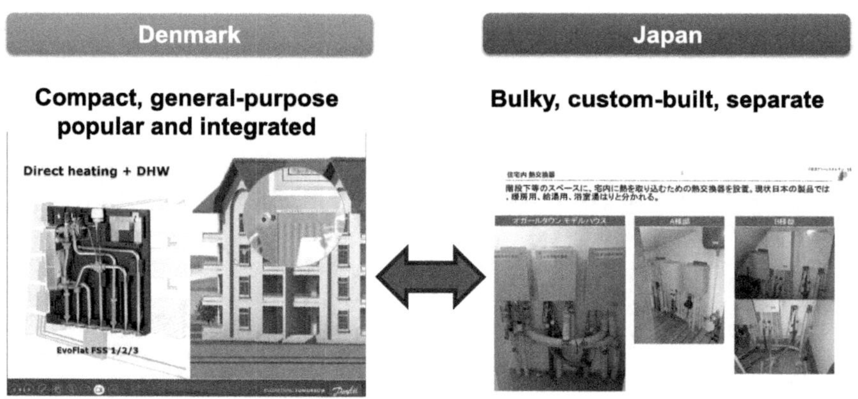

Figure 8.12 The difference in available domestic heat exchangers between Denmark and Japan

Source: Danfoss (Denmark) and Shiwa Green Energy (Japan).

Danish technology as both countries have many technical gaps. For example, large-diameter preinstalled heat pipes are not available in Japan.

Moreover, Japan has no commercially available portable heat exchangers for domestic use. In European countries such as Denmark, Austria, and Germany, you can easily find compact and integrated units that provide both room heating and hot water supply (Figure 8.12). Unfortunately, these equivalents are unavailable in the Japanese market, so if you need one, you must place a special order. The Japanese systems often come with bulky, separate units, which may be costly and unsuitable for relatively small Japanese houses.

7.3 Social and Cultural Constraints

The choice of domestic heat exchanger involves cultural aspects as well. Many Japanese households have a bath with a reheating mechanism, which is unique to Japan. In contrast, the European heat exchanger does not have this function. The choices are either going by a bulkier and more costly Japanese model or localising the European one.

Moreover, unlike Denmark, where centralised or whole-house heating systems are prevalent, Japan does not widely adopt central heating systems, except in the northernmost prefecture of Hokkaido, where over 70% of households rely on the centralised heating system. Besides, the Japanese standard for house insulation, a precondition for the centralised system (Fujiwara *et al.*, 2008), is still low compared to the European standard. Therefore, the consumer side requires significant renovation and associated costs to install the centralised system in most parts of Japan. Even in Hokkaido, spot-heating systems were taking over the centralised system because of the increased efficiency of heat pumps (Sonoda *et al.*, 1981).

8. Prospects

Despite the challenges, introducing district heating systems that use rice husk or other locally available natural resources in Japanese cities is worthwhile for the following reasons and priorities.

8.1 Advantages and Benefits of District Heating

Given the difficulties and constraints mentioned above, district heating could still be a feasible policy option since it offers several advantages and benefits for communities and individuals:

1. Local Resource Utilisation: District heating leverages natural resources available locally, contributing to regional development.
2. Cost-Effective Infrastructure Updates: Even as times change, simply changing the heat source in the plant allows cost-effective facility updates, as seen in Denmark.
3. Mitigating Heat Shock in Ageing Societies: In an ageing society with an increasing older adult population, district heating eliminates concerns about heat shock.
4. Hygienic and Safe Living Environment: Home heating becomes unnecessary, resulting in a sanitary and safe living environment.
5. Enhancing Community Welfare and Quality of Life: District heating contributes to the well-being and quality of life for individuals and the community.
6. Economic and Environmental Advantages: As decarbonisation policies progress and industries develop, district heating gains comparative advantages in affordability, safety, comfort, and environmental friendliness.
7. Integrated smart city: As seen in the Danish example, district heating networks are expected to play a crucial role in sector coupling that gains importance with advancing green transformation and digital transformation.
8. Attracting Population and Cultural Exchange: Creating attractive cities can lead to an influx of residents and increased cultural exchange, helping counter population decline.

8.2 Prospects of Rice Husk-Based District Heating

In particular, rice husk-based district heating will be a rational choice for many local cities because of the following reasons.

8.2.1 Underutilised Rice Husk Abundance Nationwide

Rice has been a strategically important crop in Japan, so the government subsidises it relatively well. Naturally, most farmers, of whom 76% are part-time farmers, choose rice as the main crop they grow. Most harvested rice is transported to local country elevators or rice centres throughout the country for processing. Specifically, Japan has 884 country elevators and 3,419 rice centres (Wada, 2014). Rice husk is often underutilised and accumulates in centres unless disposed of at a cost. Unfortunately, this practice can lead to environmental issues, as exemplified by the case in Ogata Village. Japan produces approximately 1.55 million tons of rice husk annually. The consistent supply of this substantial amount in one location makes it an ideal fuel for district heating systems.

8.2.2 Depletion of Silica from the Japanese Rice Paddy

Silica is an essential nutrient for rice. It is a primary building block of a rice plant to make its stems and husks hard. Historically, silica was replenished mainly by ploughing back rice husk charcoal produced on-site to the paddy. Nowadays, this practice is mostly banned due to environmental concerns. As a result, rice paddy became silica deficient in many parts of the country. Ogata Village has enough silica supply from silica-rich irrigation water, but few others are lucky. For instance, in neighbouring Yamagata Prefecture, silicic acid concentration in irrigation water declined by half in 40 years, from 20 ppm in 1956 to 10 ppm in 1996 (Kumagai *et al.*, 1998). According to the National Federation of Agricultural Cooperative Associations (JA), 91.8% of rice paddies nationwide suffer from silicic acid deficiency. Recycling rice husk ash or charcoal from the district heating plant to the rice paddy can effectively address the problem.

8.2.3 Localised Circular Economy and Revitalisation of Local Communities

This model enhances municipal budgets and the local economy while fostering a circular economy (Sokołowski, 2021). In Ogata Village, the annual expenditure on external energy purchases amounts to a staggering 1.1 billion JPY (Table 8.5). It represents 21% to 34% of the village's yearly budget, depending on the fiscal year (Ogata Village Office, 2023). By adopting district heating for room heating and hot water supply, much of this expenditure can be reduced. Additionally, utilising rice husk ash or charcoal, a byproduct, further enhances a sustainable circular economy. Other municipalities like Ogata Village can significantly improve their financial sustainability and contribute to a more sustainable future by producing their own renewable energy and recycling resources.

Table 8.5 Annual energy import in Ogata Village (JPY/year)

City gas and LPG	0.043
Petroleum	0.756
Electricity	0.299
Total	1.098

Source: Toshihiko Nakata Laboratory (2024).

9. Conclusion

Denmark is still leading in district heating areas, and it has shifted the central issue to smart energy systems. In 2015, Aalborg University initiated an annual international conference, smart energy systems, and Fourth Generation District Heating. Initially, the emphasis was on the fourth generation district heating, but over the years, the focus has shifted to smart energy systems, and today, more research is published on the latter.

The smart energy system concept is crucial for cost-effective 100% renewable energy systems. It emphasises energy efficiency, end-use savings, and sector integration to achieve flexibility and lower energy storage costs. Unlike the smart grid concept, which focuses solely on electricity, smart energy systems consider the entire energy system. Combining electricity, heating, cooling, and transport sectors is essential for effectively integrating fluctuating renewable sources. Energy savings and fourth-generation district heating can also be combined to create significant benefits, especially in urban areas. Electrification and district heating are pivotal in eliminating fossil fuels and ensuring renewable energy integration across sectors (Smart Energy Systems, 2024).

Incorporating smart energy systems into local Japanese cities enhances their value. It is crucial not only for the current population's welfare and well-being but also for future generations. District heating systems can play a central role in this integrated approach.

Notes

1 Heat consumption is the sum of LNG, LPG, and kerosene consumption in GJ. Kerosene consumption stands out in the Hokkaido and Tohoku regions at 59% and 44%, respectively.
2 * Converted by 1 EUR = 130 JPY, ** Residential only.
3 The line across the village's centre indicates the primary heat pipe network, which spans 3.5 km one-way (7 km for forward and return). Boxes along the heat pipe network mark the planned heat-supplied facilities.

References

Albertslund Kommune (2019) Albertslund, Copenhagen, Albertslund Kommune.

Chittum, A. and Østergaard, P.A. (2014) "How Danish Communal Heat Planning Empowers Municipalities and Benefits Individual Consumers", *Energy Policy*, 74, pp. 465–474.

Connolly, D., Lund, H., Mathiesen, B.V., Möller, B., Østergaard, P.A., Nielsen, S., Werner, S., Persson, U. and Trier, D. (2013) "The Role of District Heating in Decarbonising the EU Energy System and a Comparison With Existing Strategies", in *8th Conference on Sustainable Development of Energy, Water and Environment Systems. 8th SDEWES*, 22–27 September 2013, Dubrovnik.

Danfoss (2023) "The World's Largest Untapped Energy Source: Excess Heat". Available at: https://cdn.sanity.io/files/5zabm86v/production/88eee1dad0c4281aa8313 44903ae9a5b592be54e.pdf (Accessed: 30 May 2024).

Dochev, I., Peters, I., Seller, H. and Schuchardt, G.K. (2018) "Analysing District Heating Potential With Linear Heat Density: A Case Study From Hamburg", *Energy Procedia*, 149, pp. 410–419.

European Network on Silica (2020) "Good Practice Guide on Workers Health Protection Through the Good Handling and Use of Crystalline Silica and Products Containing It". Available at: https://guide.nepsi.eu/wp-content/uploads/2021/08/NEPSI-Good-Practice-Guide.-revised-0821pdf.pdf (Accessed: 30 May 2024).

Finnish Energy Industries (2010) "Energy Taxation in Europe, Japan and the United States". Available at: https://dokumen.tips/documents/energy-taxation-in-europe-japan-and-the-united-states-of-the-energy-taxation-survey.html?page=1 (Accessed: 30 May 2024).

Fourth Generation District Heating Forum and Institute for Sustainable Energy Policies (2020) "第4世代地域熱供給: 4DHガイドブック" [*Fourth Generation District Heating: 4DH Guidebook*]. Available at: https://www.isep.or.jp/4dh-forum/4dh-guidebook (Accessed: 30 May 2024).

Fujiwara, Y., Enai, M., Suzuki, K. and Hayama, H. (2008) "北海道の高断熱・高気密住宅におけるセントラル暖房システムの運転実態に関する調査" [*Survey on the Actual Operation and the Performance of Central Heating System for Highly Insulated and Airtight Houses in Hokkaido*], *Journal of Environmental Engineering*, 73(628), pp. 767–774.

Furubayashi, T. and Nakata, T. (2021) "導管熱密度に基づく地域熱供給ネットワークの設計手法の構築" [*Design of a District Heating Network Based on The Linear Heat Density*], *Transactions of the JSME*, 87(895).

Government of Japan (2023) "Carbon Neutrality". Available at: https://www.japan.go.jp/global_issues/carbon_neutrality/index.html (Accessed: 30 May 2024).

IEA (2013) "District Heating". Available at: https://www.iea-etsap.org/E-TechDS/PDF/E16_DistrHeat_EA_Final_Jan2013_GSOK.pdf (Accessed: 30 May 2024).

Iida, T. (ed) (2014) コミュニティパワー エネルギーで地域を豊かにする [*Community Power: Enriching Communities With Energy*]. Kyoto: Gakugei Shuppansha.

Institute for Sustainable Energy Policies 環境エネルギー政策研究所 (2021) "地域冷暖房における再生可能エネルギー導入促進のための調査委託報告書" [*Report on Commissioned Study to Promote the Introduction of Renewable Energy in District Heating and Cooling*].

IRENA Coalition for Action (2020) *Stimulating Investment in Community Energy: Broadening the Ownership of Renewables*. Abu Dhabi: International Renewable Energy Agency.

IRENA, IEA and REN21 (2020) "Renewable Energy Policies in a Time of Transition: Heating and Cooling". Available at: https://www.irena.org/-/media/Files/IRENA/Agency/Publication/2020/Nov/IRENA_IEA_REN21_Policies_Heating_Cooling_2020.pdf (Accessed: 30 May 2024).

Japan Heat Supply Business Association (2021) 熱供給事業便覧 [*Handbook of Heat Supply Businesses*]. Tokyo: Japan Heat Supply Association.

Johansen, K. and Werner, S. (2022) "Something is Sustainable in the State of Denmark: A Review of the Danish District Heating Sector", *Renewable and Sustainable Energy Reviews*, 158, p. 112117.

Kaido, K. (2001) "コンパクトシティの欧州モデルについて:持続可能な都市形態論とわが国における動き" [*The European Model of a Compact City: The Debate of Urban Form in Europe and Circumstance in Japan*], *Journal of the Real Estate Institute of Japan*, 15(3), pp. 8–17.

Kommunem, A. (2019) *Albertslund*, 9 September 2019.

Koshikawa, T., Morimoto, E. and Taniguchi, M. (2017) "Separation Between Description and Evaluation of Compact City Policy", *Journal of the City Planning Institute of Japan*, 52(3), pp. 1130–1136.

Kumagai, K., Konno, Y., Kuroda, J. and Ueno, M. (1998) "山形県における農業用水のケイ酸濃度" [*Silicic Acid Concentration in Agricultural Water in Yamagata Prefecture*], *Japanese Journal of Soil Science and Plant Nutrition*, 69(6), pp. 636–637.

Masuda, H. (2014) 地方消滅: 東京一極集中が招く人口急減 [*Regional Disappearance: Rapid Population Decline Caused by Tokyo's Concentration*]. Tokyo: Chuokoron Shinsha.
Ministry of the Environment (2019) "FY2019 Statistical Survey of CO_2 Emissions From the Household Sector (Factual Report)". Available at: https://www.env.go.jp/earth/ondanka/ghg/kateico2tokei/31co2.html (Accessed: 30 May 2024).
Ministry of the Environment (2021) "脱炭素先行地域交付金概要" [*Overview of the Decarbonisation Leading Regional Grants*].
Ministry of Foreign Affairs of Japan (2021) "Intended Nationally Determined Contributions (INDC): Greenhouse Gas Emission Reduction Target in FY2030". Available at: https://www.mofa.go.jp/ic/ch/page1we_000104.html (Accessed: 30 May 2024).
Nakata T. and Ozaki, T. (2019) "国内での地域熱供給普及の可能性" [*Possibility of Widespread District Heating System in Japan*], in: 第4世代地域熱供給4DHセミナーin東京. 主婦会館プラザエフ, p. 12.
National Institute of Population and Social Security Research (2024) "日本の地域別将来推計人口(令和5(2023)年推計)" [*The Projected Future Population of Japan by Region (2023 Estimates)*]. Available at: https://www.ipss.go.jp/pp-shicyoson/j/shicyoson23/t-page.asp (Accessed: 30 May 2024).
Nishiura, I. (2019) "小規模自治体におけるコンパクトシティ政策の課題と今後の展望" [*Issues and Future Prospects of Compact City Policy in Small Cities*], *Sapporo Otani University Faculty of Sociology Thesis Collection*, 7, pp. 1–28.
OECD (2012) "Linking Renewable Energy to Rural Development". Available at: https://www.oecd-ilibrary.org/linking-renewable-energy-to-rural-development_5k94hdrcvsr6.pdf?itemId=%2Fcontent%2Fpublication%2F9789264180444-en&mimeType=pdf (Accessed: 30 May 2024).
Ogata Village Office (2023) "令和3年度　財政状況資料集' 行政情報の公表" [*Financial Situation Data Book 2021. Publication of Administrative Information*]. Available at: https://www.vill.ogata.akita.jp/archive/contents-106 (Accessed: 30 May 2024).
Sifnaios, I., Gauthier, G., Trier, D., Fan, J. and Jensen, A.R. (2023) "Dronninglund Water Pit Thermal Energy Storage Dataset", *Solar Energy*, 251, pp. 68–76.
Smart Energy Systems (2024) "The Smart Energy System Concept". Available at: https://smartenergysystems.eu/about/ (Accessed: 30 May 2024).
Sokołowski, M.M. (2020) *European Law on Combined Heat and Power*. London: Routledge.
Sokołowski, M.M. (2021) "Models of Energy Communities in Japan (Enekomi): Regulatory Solutions From the European Union (Rescoms and Citencoms)", *European Energy and Environmental Law Review*, 30, pp. 149–159.
Sokołowski, M.M. (2022) *Energy Transition of the Electricity Sectors in the European Union and Japan: Regulatory Models and Legislative Solutions*. Cham: Palgrave Macmillan Cham.
Sonoda, T., Taniguchi, H., Tanaka, T. and Hayasaka, H. (1981) "暖房用ストーブの燃焼性能に関する研究(第1報): ポット式灯油ストーブの燃焼実験及び流動解析" [*Study on the Combustion Performance of Stoves for Space Heating (1st Report): Combustion Experiments and Analysis of the Vaporising Pot-Type Kerosene Oil-Fired Stove*], *Bulletin of the Faculty of Engineering*, 106, pp. 9–20.
State of Green (2020) "District Energy: Green Heating and Cooling for Urban Areas". Available at: https://stateofgreen.com/en/publications/district-energy/#publication-download (Accessed: 30 May 2024).
Suwa, A. (2020) "Local Government and Technological Innovation: Lessons From a Case Study of 'Yokohama Smart City Project'", in T.M. Vinod Kumar (ed) *Smart Environment for Smart Cities*. Singapore: Springer Singapore, pp. 387–403.
Tian, Z., Zhang, S., Deng, J., Fan, J., Huang, J., Kong, W., Perers, B. and Furbo, S. (2019) "Large-Scale Solar District Heating Plants in Danish Smart Thermal Grid:

Developments and Recent Trends", *Energy Conversion and Management*, 189, pp. 67–80.

Toshihiko Nakata Laboratory (2024) "地域エネルギー需給データベース(Version 2.9)" [*Japan Energy Database (Version 2.9)*]. Available at: https://energy-sustainability.jp (Accessed: 30 May 2024).

VEKS and Høje Taastrup Fjernvarme a.m.b.a. (2023) "Heat Pit Storage Optimises District Heating". Available at: https://www.veks.dk/en/publications (Accessed: 30 May 2024).

Wada, S. (2014) "農産加工·流通用機械—カントリーエレベーターの標準仕様の変遷" [*Machinery for Agricultural Products Processing and Distribution: The Changes of Country Elevator Standard Specifications*], *Journal of the Japanese Society of Agricultural Machinery and Food Engineers*, 76(6), pp. 474–479.

9 Satellite Applications for Sustainable Urban Energy

Damian M. Bielicki

1. Introduction

The global focus towards sustainable energy transition is a response to the increasing challenges posed by climate change and the growing demand for cleaner energy. Satellite technology is a powerful tool in the transition towards sustainable energy sources. Satellites also facilitate global collaboration in climate change mitigation.

This chapter examines the application of satellite data to the energy sector. It explains the role of satellites in supporting energy transition from non-renewable sources to hydropower, offshore wind projects, geothermal energy, and space-based solar panels. It also examines the role of satellites as a tool in humanity's quest to comprehend and combat the impacts of climate change. Numerous case studies from around the world are provided in support to this analysis but a particular focus is given to Japan.

2. Satellite Applications in Renewable Energy Resources on Earth

Since many states committed to the aim of achieving net zero by 2050, in recent years there has been a significant increase in investment in renewable energy in many places around the world. Some of the key energy players on the market started using satellite technology to help them with this task. Satellite applications are used in three major areas. The first area is the energy supply, which includes mapping renewable resource potential to support infrastructure siting, development, and maintenance. The second is the application for energy demand and it includes assessment of energy use patterns to predict future needs and identification of locations with unserved demand. The third is the application for energy impacts, which include monitoring the effects of energy use on climate, air quality, and water and land systems, as well as efforts to reduce these impacts (Davis, 2007, p. 202). Satellite data proved extremely useful in in these areas for renewable energy sources, such as hydropower, offshore wind projects, and geothermal power. In particular, satellite data on land use, surface topography, and roughness are extremely valuable for these

DOI: 10.4324/9781003471448-9

projects. The focus of this part is on general application of satellites to these projects, and on specific cases studies on technological developments in this area.

2.1 Offshore Wind Projects

Offshore wind power is generated through wind farms in bodies of water, usually at sea. There are higher wind speeds offshore than on land, so offshore farms generate more electricity. There are several factors to consider when selecting a site for wind projects, including wind speeds, weather impacts, topography, and many more. One of the most common satellite instruments used for such endeavours is synthetic aperture radar (SAR). It uses the energy of a sensor and receives energy as a reflection after interacting with the Earth. SAR data is used to estimate wind power from wave heights and direction, and to predict wind energy production. Wind speed is particularly critical factor in the performance of wind turbines. The higher the wind speed, the more electricity a turbine can generate. Satellites can provide both current and historical data on wind speeds to identify areas with high wind energy potential. Moreover, they provide a mean for getting situational awareness when needed by delivering near-real-time detailed imagery (Edwards *et al.*, 2022).

In practice, this technology was used to investigate the wind energy offshore Iceland. Approximately 2,500 satellite images taken over a period of seven years from European Space Agency (ESA)'s Envisat satellite helped calculate the wind speed, direction, and other factors (Nawri *et al.*, 2014). As of 2024, the Icelandic government still has not taken steps to produce offshore wind power. One of the key reasons is species protection, particularly birds and marine life, including whales. For example, long-term data satellite data drawn on measurements from ESA's Envisat Heritage Mission, as well as Copernicus Sentinel-1, revealed that wind farms can affect local air currents. However, the data can be used to locate the best locations for the future offshore wind farms to minimise the impact on natural environments (Owda and Badger, 2022).

Another major example where satellite technology was used is the data from the ERS-2 satellite for the Danish Horns Rev site in the North Sea. Horns Rev 1 was the first large-scale offshore wind farm in the world. For the project, satellite remote sensing data was used for the optimal siting, building, and operation (Hasager *et al.*, 2004). At present, satellites continue to provide valuable data, including on sand bars, which allows monitoring and estimation of the seabed dynamics, such as movement rates of the large scale sand waves (DHI Group, 2024).

In Japan, the SAR was used by a Japan-based Synspective company to estimate offshore wind speed in the Sea of Japan which borders Akita and Yamagata prefectures (Geospatial Media and Communications, 2023). Once analysed, the data will help with the decision-making process on whether a new site is suitable for the regions. It is worth noting that, as of 2024, Japan has a relatively low level of wind projects deployment, with only one large-scale offshore project at Noshiro Port in Akita Prefecture delivering approximately 140 MW, and smaller projects delivering approximately 91 MW (Zero Carbon Analytics, 2023). However, the government has increased its commitment to this technology, aiming to have up to 45 GW operational by 2040 (Reuters, 2023).

2.2 Hydropower

Hydropower is one of the oldest and largest sources of renewable energy, which uses the natural flow of moving water to generate electricity. It uses the elevation difference, created by a dam or diversion structure, of water flowing in on one side and out, far below, on the other (Office of Energy Efficiency and Renewable Energy, 2024). Images from Earth observation satellites can provide water resource assessments and thus help find locations with high energy generation potential where the environmental impacts caused by rerouting or damming the water flow would be minimal (Edwards *et al.*, 2022). Moreover, satellite data can provide information for hydrological and hydrometeorological forecast modelling, including runoff estimates, precipitation and evaporation, soil moisture, river discharge and snow-related insights. In addition, snow is an important factor, and satellites can provide an important meteorological data, as well as information about snow coverage and thickness, ground temperature, and more (ESA, 2023).

One example of a very successful space-related projects is Hypos. It is an EU-sponsored application providing a data analysis integrating Earth observation images from satellites. It includes a daily coverage for larger water bodies by using commercial satellite data. Moreover, it has a water monitoring solution providing information on chlorophyll-a, an algae bloom indicator, sea surface temperature, Secchi depth or visibility, and other data. The project has been successfully used for sites across Switzerland, Albania, and Georgia (International Hydropower Association, 2021).

Japanese hydroelectric power is developed and can be traced back to 1891 when the first hydroelectric power station commented operations in Kyoto. There are currently over 2,200 hydroelectric power stations in Japan, so this market is well developed (Tepco: Hydroelectric Power, 2023). However, there is still room for implementation of space technology, even in such an advanced infrastructure. In particular, in October 2023, it has been announced that Japan established a strategic cooperation with SpaceX's Starlink constellation of broadband satellites. These satellites will provide high-speed internet communication to hydroelectric power stations throughout Japan, mainly in mountainous regions and islands with no fibre optic or cellular access. The Suezawa Hydroelectric Power Station in Niigata Prefecture is the first station embracing this technology (J-Power: News Release, 2023).

2.3 Geothermal Energy

Geothermal energy sites, such as hot springs, fumaroles, and geysers, often exist in remote areas, making them difficult, expensive, and time-consuming to explore and measure. Therefore, the use of satellite imagery to identify geothermal resources is becoming increasingly important. They can provide a significant shortcut during the process of finding new sites, as well as monitoring of existing sites, and they are used as an addition to ground surveys and seismic measurements.

Traditional exploration ground and aerial methods are directed to finding the best suitable target locations for steam or fluid production. These are followed by hydrogeological and geochemical surveys, and a multitude of geophysical techniques necessary for geothermal characterisation (van der Meer *et al.*, 2014, p. 257). Remote sensing from satellites is capable of measuring several variables through images, gradiometry, infrared radiation, and other tools. These include mineralogy changes, direct evidence through surface features, and variations in surface temperature (van der Meer *et al.*, 2014; Nash *et al.*, 2004). In terms of monitoring, there are many ground-based geophysical and geochemical systems in and around geothermal power plants. However, such monitoring is expensive, time-consuming, and also difficult to carry out because the areas may be within national parks and may be restricted because of fumaroles or unstable regions (Mia *et al.*, 2018, p. 2).

One of the most successful satellite imagery enterprises to date is the Landsat programme, which is a joint mission of NASA and the U.S. Geological Survey (USGS), running continuously since 1972 (Landsat, 2023). The USGS uses Landsat data in the Southwest of the United States to produce mineral classification maps that indicate the presence of geothermal systems and have been used by energy companies (Landsat and Energy, 2012). Landsat satellites also discovered the pattern of geothermal enrichment at the Tengchong district in Yunnan, China, and it has become a long-term strategy for energy development in the region (Qin *et al.*, 2011, p. 552). Data from the same satellites have been used by Japanese researchers to study possible monitoring of thermal activity in Kyushu island (Mia *et al.*, 2014, 2018) and in Hokkaido island (Tian and Koike, 2014).

A successful case study from Europe is the Gravity Field and Steady-State Ocean Circulation Explorer (GOCE) mission. It was an ESA satellite operating in low Earth orbit between March 2009 and October 2013, and was used by both the ESA and the International Renewable Energy Agency. The mission was able to provide gravity maps that gave insights into particular structures favouring geothermal energy, as well as detailed crustal thickness (ESA: GOCE, 2015).

3. Space-Based Solar Power (SBSP)

SBSP is a concept first suggested by Czechoslovakian-born NASA engineer Peter Glaser in 1968 (ESA: SBSP History). In recent years, much has been done in developing SBSP, which appears to have strong benefits as a complementary energy source alongside terrestrial renewable energies. When deployed at scale, SBSP could provide environmentally sustainable, clean, affordable, continuous, abundant, and secure energy (Pelton and Madry, 2023, p. 86). This would, however, involve the space launch of multiple satellites into orbit, equipped with solar panels, converting the generated electricity into microwaves that are then transmitted wirelessly to the ground (ESA Solaris, 2023). Research and development on SBSP has been undertaken in different parts of the world. However, there are two leading countries actively working in this field: Japan and the United States.

Japan Aerospace Exploration Agency (JAXA) has had a programme of research on SBSP for many years. The Cabinet Office of the Government of Japan has included SBSP for the first time in its Basic Plan on Space Policy in 2009. The programme has also been included in the Space Plan in the following years (Japan's Basic Space Plan, 2022). In 2015, Japanese scientists at JAXA were the first to have succeeded in transmitting energy wirelessly using microwaves in a laboratory setting (JAXA RDD, 2015). In May 2023, as reported by Nikkei, the project led by Naoki Shinohara from Kyoto University will attempt to beam solar power from space as early as 2025 (Nikkei Asia, 2023). He is leading the mission, part of a project called Ohisama (Japanese for "sun"), which plans on launching a small 180-kg satellite that will transfer about 1 kilowatt of power from the altitude of 400 km (250 miles). This is about the amount of power needed to run a household appliance, such as a small dishwasher, for about an hour (Space.com, 2024).

However, it is worth mentioning that in the USA, the California Institute of Technology (Caltech) engaged in a privately funded Space Solar Power Project (SSPP) satellite that has already proven the concept. On 3 January 2023, Caltech's Space Solar Power Demonstrator (SSPD-1) satellite was flown to orbit. On 3 March 2023, for the first time in history, the SSPD-1 wirelessly beamed a detectable amount of solar energy from space towards Earth (Caltech, 2023). Even though solar power beamed from space is still a future prospect, the mission demonstrated that it is an achievable future.

4. Advantages and Challenges of Satellite Technology

Traditionally, ground and aerial data and imagery were used by most industry players. However, these are time-consuming, infrequent, not always cost-effective, and they mostly require manual analysis. The major advantage of data from space is that it can be gathered and analysed more efficiently, it is more reliable, and more sustainable. In particular, space-based solutions provide frequent data and detect most recent events. Moreover, oil and gas companies use satellite communication for health and safety reason. Satellite communication has played a major role by enabling e-training on safety and other aspects. It also enabled telemedicine to deliver healthcare in remote extraction sites (ESA, 2023c).

In terms of renewable energy resources, the examples covered in this chapter identified some of the key advantages of space technologies, including satellites being able to identify most promising locations for renewables, measuring and mapping wind speeds, water flows, temperatures, salinity gradients, minimising the environmental impact, and much more. Moreover, satellites provide effective asset management and operational risk management that help limiting the environmental impact. They allow to compare both real-time and historic data, which is especially relevant in remote areas and regions with limited funding and under-developed infrastructure (ESA, 2023a).

Despite numerous advantages that space applications bring, there are also several challenges, which are mostly associated with operation of satellites in space. The first concern is about cybersecurity. Satellites supporting energy sector have already been subject to hacking. In particular, in February 2022, a satellite outage the affected operation – monitoring and control – of over 5,000 wind turbines across Europe (EuroNews, 2022). Consequently, space-based systems, as well as ground systems, must be adequately protected from any form of interference (Freeland and Martin, 2023, pp. 281–289; see also Bonnart *et al.*, 2023, p. 49). Second, there is the issue of the number of slots available on certain orbits, particularly on the geostationary orbit. This could be an issue for projects such as SBSPs, which are likely to utilise this orbit. The International Telecommunication Union (ITU) highlighted that orbits and associated radio frequencies are limited natural resources and "must be used rationally, efficiently and economically" (Article 44, ITU Convention; see also Lyall, 2015). Third, satellites in any Earth orbit are exposed to space weather. Solar storms, in particular coronal mass ejections, do occur and they can disrupt power grids and damage satellites (NOAA, 2023). Space debris is yet another threat. Broadly speaking, space debris are non-functional man-made objects that remain in space. They are a grand challenge because millions of small, undetectable, but harmful pieces of debris are present in different Earth orbits (Byers and Boley, 2023, pp. 7–8).

All these examples present major challenges for all satellite technologies. However, specifically in relation to the SBSP, there is also the issue of transmission of the energy to Earth. SBSP has its obvious advantages, including provision of a renewable energy source 24 hours a day and under any weather conditions. Moreover, power could be directed to different locations without the need for power lines. In practice, the SBSP would convert transmit power down to ground-based rectennas, and these require large areas for receiving the energy. One of many reasons for it is the fact that overly concentrated microwave bean could potentially affect animals, birds, people, and aircrafts in flight (Pelton and Madry, 2023, p. 87). The SBSP is a promising technology but, at the time of writing, it is not yet operational, and it is likely to take years before it is implemented on a large scale. In 2024, NASA has released a report questioning the feasibility of the technology (NASA SBSP, 2024). The project would require the development of a constellation of satellites, which is a costly technology. Moreover, it may be difficult (if not impossible) to maintain/repair the satellites in space and so they would need to be replaced in case of damage or near their end-life. The average lifetime of such satellites is about 15 years but there are economic and environmental pressures to extend the life of all space assets (JAXA: SSPS, 2023). Finally, the effectiveness of SBSP depends on position of the satellites in space. When located in geosynchronous orbit, satellites are rarely blocked from solar radiation. This would allow energy to be continuously beamed down to Earth. However, as stated above, this orbit is a limited natural resource (Pelton and Madry, 2023, pp. 86–87).

5. Satellites and Climate Change Mitigation

The core reason for transition to sustainable energy sources is the impact of non-renewable energy sources on climate change. In accordance with the Paris Climate Agreement commitments, the States agreed to achieve net zero by 2050 (Paris Agreement, 2015). The Agreement is a roadmap with renewable energies being on the top of the agenda. It is also in line with the UN Sustainable Development Goals, which focus on climate action, responsible consumption and production, sustainable cities and communities, affordable and clean energy, and good health and well-being among others. Moreover, a crucial link, facilitated by ICT, exists between sustainable energy and smart cities, which is bound to lead to breakthroughs in the energy sector (Sokołowski and Visvizi, 2023, pp. 1–9). For example, Yokohama and Kanagawa Prefectures are places where numerous technology-related companies have their sites, including Panasonic, Honda, and Apple. They generate a lot of data and, consequently, energy. Therefore, their reliance on sustainable energy sources is growing, as there is an increasing pressure on large companies to adopt climate-friendly policies. This correlation is further enhanced by data from the space industry (Regan, 2022).

With the Paris Climate Agreement and the UN Sustainable Development Goals in mind, Japan revised its Basic Plan on Space Policy in 2020 by focusing on sustainable use of outer space and implementation of the UN Sustainable Development Goals. The policy had a major impact on the national space agency, JAXA, with its focus shifting towards scientific research and technological development supporting the understanding of climate change and sustainability. Implementation of this policy requires investment in and implementation of satellite infrastructures (Japan's Basic Space Plan, 2022).

The role of space technology in supporting the two major global agendas and energy transition is increasing steadily. Traditionally, satellites supported major energy providers in monitoring and controlling their environmental impact. In particular, satellites provide monitoring of infrastructure to avoid leaks or spills and gas flaring. Moreover, satellites monitor greenhouse gases emissions, such as methane and carbon dioxide, which have major environmental impact. Other types of monitoring and measurement include precipitation, sea ice, sea surface temperature, and soil moisture (ESA, 2023c, p. 3).

A good example of risk mitigation projects that has an impact on the environment is the "Corridor and Asset Monitoring using Earth Observation". It is an ESA project offering satellite-derived information to monitor energy grid and prevent leakages and outages. For instance, it provides vegetation monitoring along pipeline, particularly in forested areas. It also provides a very precise monitoring of major events, such as floods, and geospatial information to support emergency management activities (Cameo, 2023). ESA also supported other projects through its Business Applications and Space Solutions (BASS) programme. One particularly interesting project supported by BASS is the Methane Watch by a French company Kayrros. It uses data from Sentinel-5 satellite to monitor methane super-emitters in near real time. It helps governments and businesses understand the risks posed by the changing climate and energy landscapes and make more informed decisions. The technology is also being used by the UN Environment Programme (Kayrros Methane Watch, 2023).

Another relevant project is the World Bank's Global Gas Flaring Tracker, which has been recognised as the only global and independent indicator of routine gas flaring. The tracker uses data from satellites operated by the U.S. National Oceanic and Atmospheric Administration (NOAA) which allow monitoring of global flaring levels and track progress towards their goal of Zero Routine Flaring by 2030 (World Bank: GGFR, 2023). Moreover, a London-based Capterio company created in 2021 a digital tool called FlareIntel, which every day tracks every flare for every asset, for every company, in every country. The project uses satellite data from Earth Observation Group, specialising in nighttime observations of lights and combustion sources (FlareIntel, 2023).

There are a few satellite-related projects in Japan that significantly contribute to the global climate change mitigation. In 2009, Japan has become the second country in the world, after the United States, that monitors greenhouse gas emissions through its own satellites (Schrogl, 2020, p. 1383). It operates the Greenhouse Gases Observing Satellites (GOSAT – known in Japan as *Ibuki*). They measure the densities of carbon dioxide and methane through thermal and near infrared sensors for carbon observation, i.e. Fourier transform spectrometer and cloud and aerosol imager. The results contribute to the climate science and rare useful for climate change-related policies (GOSAT, 2023). Since 2012, Japan has also been monitoring precipitation, sea ice, sea surface temperature, and soil moisture, through its two satellites operating for the Global Change Observation Mission (GCOM). The satellites, known as *Shikisai* and *Shizuku*, use advanced microwave scanning radiometers that help to enhance the accuracy of forecasting environmental changes (JAXA: GCOM, 2023). One more interesting venture is the JJ-FAST programme, which is a satellite monitoring system for deforestation in tropical regions. It uses JAXA's Advanced Land Observing Satellite (ALOS-2) with synthetic aperture radar to record changes on the Earth's surface, such as illegal logging and climate change. The data is uploaded to a website where anyone can freely access forest-monitoring maps (JICA-JAXA, 2023). The project contributes not only to forest management, but also biodiversity protection and, ultimately, to the fight against global warming.

6. Conclusions

Since the first satellite images were made publicly available in 1972, applications of satellite data have expanded significantly (Davis, 2007). Satellite data offers the possibility of synergies between the different energy markets in different parts of the world. Nowadays, the predominant role of satellite technology in the energy sector is not to supply the energy to Earth, but to provide shortcuts in determining the right locations for new sites, and to make existing energy systems more efficient and secure. Satellite data complement other information sources to provide a more complete picture, often with continuous spatial coverage (Edwards *et al.*, 2022). Moreover, the role of Earth observation satellites in climate mitigation is well established. The measurements from space missions provide an invaluable data concerning climate monitoring, as they generate consistent, long-term, and global data records for the key essential climate variables that support the key international climate actions.

Space sustainability is an intrinsically multilateral aspect that is a major concern for space-related activities. It refers to the issues such as space debris, orbital slots allocation, harmful interference, and other risks that can impact the continuity of the benefits derived on Earth from space activities (Schrogl, 2020, pp. 319–320). Nevertheless, space-based technology offers a unique opportunity to enhance urban living, and plays a pivotal role in supporting sustainability, mobility, connectivity, health, resilience, and other aspects of daily life.

As of June 2024, Japan's reliance on fossil fuels stands at over 80%. In order to achieve the aim of carbon neutrality by 2050, a substantial acceleration towards sustainable sources is required. As seen in the given examples, satellite applications can significantly speed up the process. It meets needs for both sustainable urban development, disaster mitigation, and addressing the international efforts to achieve the global sustainable development goals. Japan is effectively connecting these with satellite and other technologies.

References

Bonnart S., Pelin Mantı, N., Breda, P., de Beaumont, M.R. and Jha, D. (2023) "Cybersecurity Threats to Space: From Conception to the Aftermaths", in P.J. Blount and M. Hofmann (eds) *Space Law in a Networked World*. Nijhoff: Brill, pp. 39–101.

Byers, M. and Boley, A. (2023) *Who Owns Outer Space? International Law, Astrophysics and the Sustainable Development of Space*. Cambridge: Cambridge University Press.

Caltech (2023) "In a First, Caltech's Space Solar Power Demonstrator Wirelessly Transmits Power in Space". Available at: https://www.caltech.edu/about/news/in-a-first-caltechs-space-solar-power-demonstrator-wirelessly-transmits-power-in-space (Accessed: 2 June 2024).

Cameo (2023) "Earth Observation". Available at: https://www.cameo-platform.com/ (Accessed: 2 June 2024).

Constitution and Convention of the International Telecommunication Union (ITU) Adopted by the 2022 Plenipotentiary Conference. 26 September–14 October 2022, Bucharest.

Davis, G. (2007) "History of the NOAA Satellite Program", *Journal of Applied Remote Sensing*, 1, p. 012504.

DHI Group (2024) "Satellites and Offshore Wind Farms? A Match Made in Space". Available at: https://dhigroup.com/satellites-and-offshore-wind-farms/ (Accessed: 2 June 2024).

Edwards, M.L., T. Holloway, R.B. Pierce, L. Blank, M. Broddle, E. Choi, B.N. Duncan, Esparza, Á., Falchetta, G., Fritz, M., Gibbs, H.K., Hundt, H., Lark, T.J., Leibrand, A., Liu, F., Madsen, B., Maslak, T., Pandey, B., Seto, K.C. and Stackhouse, P.W. (2022) "Satellite Data Applications for Sustainable Energy Transitions", *Frontiers in Sustainability*, 3, pp. 5–6.

ESA (2015) "GOCE Helps Tap into Sustainable Energy Resources". Available at: https://www.esa.int/Applications/Observing_the_Earth/FutureEO/GOCE/GOCE_helps_tap_into_sustainable_energy_resources (Accessed: 2 June 2024).

ESA (2023a) "How Space Supports Energy Transition". Available at: https://commercialisation.esa.int/2020/09/how-space-supports-the-energy-transition/ (Accessed: 2 June 2024).

ESA (2023b) "SBSP History". Available at: https://www.esa.int/Enabling_Support/Space_Engineering_Technology/SOLARIS/SBSP_history (Accessed: 2 June 2024).

ESA (2023c) "Space Meets the Oil and Gas Industry: Use Cases and Market Opportunities". Available at: https://commercialisation.esa.int/wp-content/uploads/2023/09/Oil_and_gas_report_230907.pdf (Accessed: 2 June 2024).

ESA (2023d):"Space Meets Hydro Energy: Use Cases and Market Opportunities". Available at: https://commercialisation.esa.int/wp-content/uploads/2023/07/Space_meets_hydro_energy-Mini-report-1.pdf (Accessed: 2 June 2024).

ESA Solaris (2023) "Cost Vs Benefits Studies". Available at: https://www.esa.int/Enabling_Support/Space_Engineering_Technology/SOLARIS/Cost_vs._benefits_studies (Accessed: 2 June 2024).

EuroNews (2022) "Satellite Outage Knocks Out Thousands of Enercon's Wind Turbines". Available at: https://www.euronews.com/next/2022/02/28/ukraine-crisis-cyber-enercon (Accessed: 2 June 2024).

FlareIntel (2023) "Track Every Gas Flare". Available at: https://flareintel.com/ (Accessed: 2 June 2024).

Freeland, S. and Martin, A.S. (2023) "Harnessing Artificial Intelligence Technologies for Sustainable Space Missions: Legal Perspectives", in M. Madi and O. Sokolova (eds) *Intelligence for Space: AI4SPACE Trends, Applications, and Perspectives*. London: Routledge, pp. 281–289.

Geospatial Media & Communications (2023) "SAR Catalyzing Offshore Wind Energy Projects". Available at: https://www.geospatialworld.net/prime/sar-catalyzing-offshore-wind-energy-projects/ (Accessed: 2 June 2024).

GOSAT (2023) "Greenhouse Gases Observing Satellite GOSAT 'IBUKI'". Available at: https://www.gosat.nies.go.jp/ (Accessed: 2 June 2024).

Hasager, C.B., Dellwik, E., Nielsen, M. and Furevik, B. (2004) "Validation of ERS-2 SAR Offshore Wind-Speed Maps in the North Sea", *International Journal of Remote Sensing*, 25(19), pp. 3817–3841.

International Hydropower Association (2021) "Satellite Images Can Now Tell You (Almost) Everything About Your Reservoir". Available at: https://www.hydropower.org/case-study/satellite-reservoir-management (Accessed: 2 June 2024).

Japan's Cabinet Office (2022) "Basic Space Plan". Available at: https://www8.cao.go.jp/space/plan/keikaku.html (Accessed: 2 June 2024).

JAXA (2023a) "About the SSPS". Available at: https://www.kenkai.jaxa.jp/eng/research/ssps/ssps-ssps.html (Accessed: 2 June 2024).

JAXA (2023b) "GCOM". Available at: https://global.jaxa.jp/projects/sat/gcom_w/ (Accessed: 2 June 2024).

JAXA RDD (2015) "Ground Demonstration Testing of Microwave Wireless Power Transmission". Available at: https://www.kenkai.jaxa.jp/eng/research/ssps/150301.html (Accessed: 2 June 2024).

JICA-JAXA (2023) "Forest Early Warning System in the Tropics". Available at: https://www.eorc.jaxa.jp/jjfast/system.html (Accessed: 2 June 2024).

J-Power: News Release (2023) "Starlink Integration at a Hydroelectric Power Station in a Mountainous Region in Japan". Available at: https://www.jpower.co.jp/english/news_release/pdf/news231026.pdf (Accessed: 2 June 2024).

Kayrros Methane Watch (2023) "Kayrros Methane Watch: Independent Monitoring of Large Methane Emissions". Available at: https://methanewatch.kayrros.com/ (Accessed: 2 June 2024).

Landsat (2023) "50 Years of Landsat Science". Available at: https://landsat.gsfc.nasa.gov/ (Accessed: 2 June 2024).

Landsat and Energy (2012) "Landsat's Critical Role in Energy". Available at: https://landsat.gsfc.nasa.gov/wp-content/uploads/2013/02/LandsatEnergyFS.pdf (Accessed: 2 June 2024).

Lyall, F. (2015) "'Harmful Interference' and the ITU", in M. Hofmann (ed) *Harmful Interference in Regulatory Perspective: Legal Rules for Interference-Free Radio Communication*. Baden-Baden: Nomos, pp. 19–29.

Mia, M.B., Fujimitsu, Y. and Nishijima, J. (2014) Exploration and Monitoring Geothermal Activity Using Landsat ETM+Images: A Case Study at Aso Volcanic Area in Japan", *Journal of Volcanology and Geothermal Research*, 275(4), pp. 14–21.

Mia, M.B., Fujimitsu, Y. and Nishijima, J. (2018) "Monitoring of Thermal Activity at the Hatchobaru-Otake Geothermal Area in Japan Using Multi-Source Satellite Images: With Comparisons of Methods, and Solar and Seasonal Effects", *Remote Sensing*, 10(9), p. 1430.

NASA SBSP (2024) "Space-Based Solar Power", Report ID 20230018600. Available at: https://www.nasa.gov/wp-content/uploads/2024/01/otps-sbsp-report-final-tagged-approved-1-8-24-tagged-v2.pdf?emrc=744da1 (Accessed: 2 June 2024).

Nash, G.D., Johnson, G.W. and Johnson, S. (2004) "Hyperspectral Detection of Geothermal System-Related Soil Mineralogy Anomalies in Dixie Valley, Nevada: A Tool for Exploration", *Geothermics*, 33(6), pp. 695–711.

Nawri, N., Petersen, G.N., Bjornsson, H., Hahmann, A.N., Jónasson, K., Hasager, C.B. and Clausen, N.E. (2014) "The Wind Energy Potential of Iceland", *Renewable Energy*, 69, pp. 290–299.

Nikkei Asia (2023) "Japan to Try Beaming Solar Power From Space in Mid-Decade". Available at: https://asia.nikkei.com/Business/Science/Japan-to-try-beaming-solar-power-from-space-in-mid-decade (Accessed: 2 June 2024).

NOAA (2023) "Space Weather Conditions: Space Weather Prediction Center". Available at: https://www.swpc.noaa.gov/communities/space-weather-enthuiasts (Accessed: 2 June 2024).

Office of Energy Efficiency and Renewable Energy (2024) "Hydropower Basics". Available at: https://www.energy.gov/eere/water/hydropower-basics (Accessed: 2 June 2024).

Owda, A. and Badger, M. (2022) "Wind Speed Variation Mapped Using SAR Before and After Commissioning of Offshore Wind Farms", *Remote Sensing*, 24(6), p. 1463.

Paris Agreement to the United Nations Framework Convention on Climate Change (2015) Adopted on 12 December 2015, Entered into Force on 4 November 2016.

Pelton, J.N. and Madry, S. (2023) "Space Systems, Quantum Computers, Big Data and Sustainability: New Tools for the United Nations Sustainable Development Goals", in M. Madi and O. Sokolova (eds) *Intelligence for Space: AI4SPACETrends, Applications, and Perspectives*. London: Routledge, pp. 53–106.

Qin, Q., Zhang, N., Nan, P. and Chai, L. (2011) "Geothermal Area Detection Using Landsat ETM+ Thermal Infrared Data and Its Mechanistic Analysis: A Case Study in Tengchong, China", *International Journal of Applied Earth Observation and Geoinformation*, 13(4), pp. 552–559.

Regan E. (2022) "Smart Cities in Japan: Practical Innovations for Conscious Future", Toquoesque. Available at: https://tokyoesque.com/smart-cities-in-japan/#:~:text=Smart%20cities%20such%20as%20Kashiwa,to%20build%20a%20smarter%20future (Accessed: 2 June 2024).

Reuters (2023) "Japan Aims to Become Major Offshore Wind Energy Producer". Available at: https://www.reuters.com/business/energy/japan-aims-become-major-offshore-wind-energy-producer-2023-06-30/ (Accessed: 2 June 2024).

Schrogl, K.-U. et al. (eds) (2020) *Handbook of Space Security: Policies, Applications and Programs*. Cham: Springer.

Sokołowski, M.M. and Visvizi, A. (2023) "Exploring the Smart Cities–Energy Communities Nexus: Challenges and Issues", in M.M. Sokołowski and A. Visvizi (eds) *Routledge Handbook of Energy Communities and Smart Cities*. London: Routledge, pp. 1–9.

Space.com (2024) "Japanese Satellite Will Beam Solar Power to Earth in 2025". Available at: https://www.space.com/japan-space-based-solar-power-demonstration-2025 (Accessed: 2 June 2024).

Tepco (2023) "Hydroelectric Power". Available at: https://www.tepco.co.jp/en/rp/ourbusiness/hydro/index-e.html#:~:text=(The%20second%20hydroelectric%20power%20station,vicinity%20of%20the%20power%20station (Accessed: 2 June 2024).

Tian, B. and Koike, K. (2014) "Mapping Spatial Variability of Geothermal Resource in Hokkaido Japan by Combination of Thermal Remote Sensing and Borehole Data", in *2014 IEEE Geoscience and Remote Sensing Symposium. 2014 IGARSS*, 13–18 July, Quebec City: IEEE, pp. 879–882.

van der Meer F., Hecker, C., Ruitenbeek, F.V., Werff, H., de Wijkerslooth, C. and Wechsler, C. (2014) "Geologic Remote Sensing for Geothermal Exploration: A Review", *International Journal of Applied Earth Observation and Geoinformation*, 3, pp. 255–269.

World Bank (2023) "Global Gas Flaring Reduction Partnership". Available at: https://www.worldbank.org/en/programs/gasflaringreduction/global-flaring-data (Accessed: 2 June 2024).

Zero Carbon Analytics (2023) "Offshore Wind in Japan: The Untapped Potential". Available at: https://zerocarbon-analytics.org/archives/energy/offshore-wind-in-japan-the-untapped-potential (Accessed: 2 June 2024).

10 Energy Transition in Japan's SDGs Future Cities

Toyama, Sapporo, and Kumamoto

Hiroshi Ito

1. Introduction

A smart city is an urban environment aiming to socially, ecologically, and economically enhance the quality of life of residents by employing information and communication technologies (ICTs) (Zuccala and Verga, 2017; Su and Fan, 2023). The term "smart city" originated in the 1990s alongside technological development (see Hollands, 2008; Söderström *et al.*, 2014), which played a significant role in shaping modern urban infrastructure, including smart mobility, buildings, and grids (Thornbush and Golubchikov, 2021). Furthermore, in 1992, the adoption of Agenda 21 at the United Nations Conference on Environment and Development in Rio de Janeiro with a focus on "smart growth" in urban areas propelled the development of smart cities (Singh and Singla, 2020).

Two decades later, the term "smart energy city" emerged when it was used in the 2010s in response to the energy-related components of smart and low-carbon cities to address urban energy demands and the effects of climate change (Thornbush and Golubchikov, 2021). In this light, smart energy cities aim for high energy efficiency, with smart transportation networks and buildings, innovative public lighting systems, and reduced air pollution and CO_2 emissions (Battarra *et al.*, 2016). These aims will be achieved by deploying Internet of Things (IoT) technologies within city infrastructure for optimally employing renewable energy, including water, wind, tidal, biofuels, geothermal, and solar energy (Somayya and Ramaswamy, 2016). Energy efficiency and conservation can be achieved further by employing ICTs. For example, sensor technology and smart metering devices provide information on timely energy consumption (Zuccala and Verga, 2017). In this context, this chapter discusses Japan's Sustainable Development Goals (SDGs) Future Cities initiative, specifically homing in on the development of strategies that tackle the challenges associated with energy.

This chapter is structured as follows. The subsequent section examines Japan's unique approaches and initiatives towards smart cities, along with an exploration of how these align with the SDGs within the SDGs Future Cities initiative. In the methodology section, the research approach is described, encompassing document analysis and semi-structured interviews. The results

DOI: 10.4324/9781003471448-10

section presents a comprehensive analysis of the strategies of Toyama, Sapporo, and Kumamoto, which are among the 182 SDGs Future Cities designated by the Japanese government and were selected for this study. Subsequently, a discussion is presented that contrasts and compares the shared and distinctive aspects of their commitment to sustainability, innovation, and resilience. The chapter concludes with a synthesis of the findings, emphasising the global implications for smart city development and highlighting the role of these cities as exemplars of sustainable urban planning.

1.1 Smart Energy Cities in Japan

Since 2016, the Japanese government has promoted a concept called Society 5.0 or "super smart society", in which various social needs are met through the provision of high-quality products and services. This initiative was designed after the European model of Industry 4.0 (St-Pierre, 2023). Society 5.0 aims for a human-centred society that seeks solutions to socio-economic issues through a system integrating cyber space (virtual space) and physical space (real space), as proposed and approved by the Cabinet Office in the Fifth Science and Technology Basic Plan. This plan was a vision for a future Japanese society for the years 2016–2020 and beyond (Government of Japan, 2016). Influenced by the recognition of specific challenges in the field of smart cities during the 2018 Davos Conference (Nakamura and Krizaj, 2020), the Revised National Strategic Special Zones Law (Act 34 of 2020), also known as the Super City Initiative, was enacted in May 2020 to develop smart city initiatives in Japan (Library of Congress, 2020). This legislation was originally implemented in 2013, and the Japanese government later adopted the term "super" to further promote smart cities (Library of Congress, 2020).

As Sato *et al.* (2022) observed, projects related to smart cities initiated by various government agencies, such as the Ministry of Economy, Trade and Industry and the Ministry of Land, Infrastructure, Transport and Tourism (MLIT), primarily focus on specific areas such as renewable energy and disaster preparedness. DeWit (2013) argued that although the definitions of Japanese smart cities might lack precision, they explicitly emphasise energy-related aspects, particularly the mitigation of carbon emissions. Such emphasis demonstrates the potential of the Japanese smart city concept, specifically its capacity to decentralise the energy sector in Japan and stimulate the revival of local economies (Deguchi, 2018).

Some unique characteristics of Japanese smart energy cities include smart community models, microgrids, smart grids, and disaster resilience (Sokołowski, 2021). In Japan, the smart community model has gained prominence because of its innovative approach to energy distribution. This model is built on a framework relying on a network of small-scale power distribution systems. In contrast to traditional, large-scale centralised energy distribution networks, the Japanese smart community model offers several distinct advantages. One key strength is its enhanced energy security, as it reduces the vulnerability associated with a single centralised power source. In addition, this approach is viewed as potentially more resilient when confronted by disruptions such as natural disasters, as it allows for distributed energy generation and management, thereby minimising the impact of localised outages. Consequently, Japan's smart community model has increasingly been recognised as a forward-thinking and robust alternative to energy distribution (Zappa, 2020).

Smart energy cities in Japan often feature microgrids and smart grid systems. Microgrids enable localised energy generation, distribution, and consumption, thereby improving the stability and resilience of the energy infrastructure. These grids can function autonomously during power outages, and they offer additional advantages, such as reductions in the carbon footprint and, in many cases, in overall energy costs (Deguchi, 2018). Smart grids are at the forefront of modern energy management and integrate advanced sensors, automation technologies, and sophisticated data analytics. The integration of ICTs seeks to attain decarbonisation objectives by incorporating a higher proportion of renewable energy sources, enhancing supply reliability through automated grid adjustments and empowering consumers to actively engage in energy markets (Quitzow and Rohde, 2022).

In view of Japan's exposure to natural disasters, the concept of smart energy cities places strong emphasis on bolstering disaster resilience. In Japan, smart energy cities have been designed to encompass a robust energy infrastructure that places a premium on resilience. This infrastructure includes a network of backup power systems, decentralised and distributed generation framework, and comprehensive disaster-response plans. By integrating these elements, smart energy cities can not only maintain a dependable energy supply during emergencies but also ensure the continuity of essential services and operations. This proactive approach to disaster resilience underscores Japan's commitment to safeguard communities and energy resources in the face of adversity.

1.2 Japan's SDGs Future Cities

In Japan, the aims of the 1997 Kyoto Protocol Target Achievement Plan, which was formulated and took effect in 2005 and subsequent relevant policies were to reduce greenhouse gas (GHG) emissions and create a low-carbon society (Sokołowski, 2022). However, in the early 2010s, Japan was confronted by significant challenges after the 2011 Great East Japan Earthquake, followed by a massive tsunami. This catastrophic event led to the re-evaluation of Japan's energy policy and the role of nuclear power and renewable energy. The country embarked on a "green energy revolution" to reduce its dependence on nuclear power as much as possible (Sokołowski, 2022). The 2014 Fourth Strategic Energy Plan of Japan emphasised the idea of "moving away from dependence on nuclear power" and shifting to renewable energy (Agency for Natural Resources and Energy, 2014). The adoption of SDGs and signing of the Paris Agreement in 2015 underscored the importance of smart energy and sustainable cities (Zappa, 2020).

Of the 17 SDGs, SDG-7 focuses on ensuring access to affordable, reliable, sustainable, and modern energy, whereas SDG-11 aims to develop inclusive, safe, resilient, and sustainable cities. The Japanese Government has taken advantage of this SDG framework. In 2018, the Cabinet's Regional Revitalisation Promotion Office solicited model projects under the SDGs Future City initiative (Cabinet of Japan, 2018; Deguchi, 2018). This initiative is supported by the national-level programme SDGs for Regional Revitalisation Public–Private Partnership Platform and selected local governments with promising initiatives aligned with the SDGs to promote good/best practices for local and regional development (Masuda *et al.*, 2022). *Seen in this light,* Japan promotes the self-generation and self-consumption of renewable energy (SDG-7), contributing to regional revitalisation and disaster prevention (Sokołowski, 2022). Regional and local policies and development plans are key tools for designing interventions and targeted measures that support the successful implementation of SDG-11 (Papadopoulou, 2021).

Between 2018 and 2023, the Japanese government designated 182 smart cities as SDGs Future Cities, with proposals for solving relevant issues and creating new value in three aspects, namely economy, society and the environment. Smart city projects often target specific areas such as renewable energy and disaster countermeasures (Sato *et al.*, 2022) in the context of achieving the SDGs (Nago *et al.*, 2021).

Although Japan trails behind in its digitalisation efforts (St-Pierre, 2023) and the West has taken the lead in adopting smart grids and related systems (Deguchi, 2018), Japan's case studies, including those of the SDGs Future Cities, could provide valuable insights and lessons for city planners and urban developers both locally and globally. These experiences work towards addressing economic, environmental, and social challenges, such as population decline and ageing through smart energy initiatives (Balaban and de Oliveira, 2022; Masuda *et al.*, 2022).

2. Methodology

A multiple case study approach was followed in this research, as it could yield significant analytical benefits, i.e. evidence obtained from multiple cases tend to enhance validity compared with that derived from a single case (Yin, 2018). Each case was selected carefully, resulting in distinct case studies with comparable results. In most case studies, the primary concern is not formal theory – instead, researchers typically work with a proposition of interest from the literature intended for further investigation (Stewart, 2012).

2.1 Sampling

This study investigates three Japanese cities, namely Toyama, Sapporo, and Kumamoto that have been designated as SDGs Future Cities. These cities were selected for the following reasons. Toyama City has been making efforts to establish itself as a smart city by utilising digital technology and data to promote its city strategies (Toyama City, 2022a). Toyama City has accomplished energy efficiency by establishing a compact city through well-coordinated public transportation policies. This success was recognised by the United Nations Sustainable Energy for All (SE4ALL) Initiative when the city was designated one of the 13 model "energy efficiency improvement cities" in 2014 (Toyama City, 2015).

Sapporo City was selected as one of 37 Smart City Model Projects by Japan's MLIT in 2019 (Matsumoto *et al.*, 2023). The aim of these model projects is solving urban and regional issues by using new technologies and big data. Sapporo City is located in northern Japan, which has a cold and snowy winter climate; therefore, energy generation and conservation have been significant local concerns. As is elucidated in subsequent sections, Sapporo has actively undertaken measures to address this challenge by employing pioneering policies and practices.

The initiatives of Kumamoto City have gained widespread recognition for creating a community that is resilient to disasters through innovative energy strategies. Informed by the experience and insights gained from the 2016 earthquakes, the city embarked on an ambitious trajectory towards the establishment of a resilient and intelligent disaster-resistant urban centre. Pivotal to this multifaceted endeavour is the role of the community energy enterprise known as Smart Energy Kumamoto, which supplies electricity generated exclusively from renewable energy sources towards the realisation of a decarbonised society.

2.2 Methods

Data collection was done through policy document analysis and semi-structured interviews. Policy document analysis serves as a research method for scrutinising policy documents (Cardno, 2018). Using this method, non-technical literature such as reports and internal correspondence emerge as empirical data for conducting case studies that focus on singular phenomena, events, organisations, or programmes (Bowen, 2009). One distinguishing advantage of document analysis is the accessibility of numerous documents without requiring author permission (Bowen, 2009). In addition, public document analysis can mitigate certain ethical considerations linked to alternative qualitative methods (Morgan, 2021). A potential drawback is that policy documents are not created specifically for research, which could result in failure to meet study purposes. Furthermore, confirming the relevance of these documents can be challenging (Cardno, 2018). In conducting the study, we accessed policy documents available through online sources.

Conducting interviews is an important data collection method for case studies. We conducted interviews to supplement the document analysis findings, develop a deeper understanding, elicit factual material, and validate perspectives (Stewart, 2012). Interviewees included 12 city government officials from relevant divisions in the municipalities of Toyama (three respondents), Sapporo (four respondents), and Kumamoto (five respondents). The interviews with Toyama City officials were conducted on 13 September 2018, 22 March 2023, and 17 April 2024 (with one official in each interview). Meanwhile, interviews with Sapporo City officials and Kumamoto City officials were held on 11 May 2023 and 19 May 2023, respectively. The names of the respondents were anonymised. The interview questions were drawn from the document analysis. The findings provide information on energy policies and practices of each of the three cities examined in this study.

2.3 Limitations of Research

This study is subject to limitations. Primarily, the research focused on a limited sample size of three cities, employing document analysis and interviews. While these methods provided valuable insights, it is crucial to acknowledge the potential selection bias. Consequently, caution must be exercised in generalising the findings beyond the specific context of the examined cities. The reliance on certain methods also imposes constraints on the study outcomes. To address these limitations, future research should broaden the number of cities included to improve generalisability and reduce potential bias. Employing diverse research methods, including quantitative approaches, offers a perspective that is more holistic and enhances data triangulation. Doing so contribute significantly to advancing the knowledge base within the field, thereby fostering a comprehensive understanding of the complexities inherent in smart energy cities.

3. Results

The research findings are detailed in the subsequent sections, starting with Toyama, a city that exemplifies the synergy between abundant natural resources and innovative urban planning. Next is Sapporo, a city that has transformed its climate challenges into opportunities, pioneering smart city projects that leverage snow cooling systems and district heating to optimise energy use. Finally, the focus shifts to Kumamoto, where a proactive stance on disaster resilience and a commitment to decarbonisation have shaped its approach to smart energy management.

3.1 Toyama

The city of Toyama, the capital of the Toyama Prefecture in Japan, is situated along the coast of the Sea of Japan in the Hokuriku region. With a land area of 1,240 km^2 (479 sq. miles) and a population of 404,044 in 2024 (a decline from its 2010 peak of 422,000), the city has been grappling with demographic shifts (Toyama City, 2024). Toyama has abundant natural resources, a notable strength of the city and the Prefecture, enabling various aspects of self-sufficiency. Toyama boasts rich energy resources, including geothermal and hydroelectric power, partly because of its extensive forested areas (covering 70% of the city). For instance, hydroelectric power generation alone covers the city's current electricity consumption (Toyama Prefecture, 2018). Such resources are instrumental in local energy production and consumption and are called *chisan chisho* ("local production for local consumption")[1] (Toyama City and IGES, 2018).

Building on its wealth of natural resources and commitment to local energy solutions, Toyama City has also ventured into international cooperation to address power shortages and sustainable energy production. In March 2014, Toyama City and Regency Tabanan in Indonesia concluded a cooperation agreement, which, with the support of the Japan International Cooperation Agency led to the launch of a significant renewable energy project. Together with business partners, Toyama City initiated research and installed four micro-hydroelectric plants in the irrigation channels of the terraced rice fields of Jatiluwih Village, a site recognised by the United Nations Educational, Scientific and Cultural Organisation for its cultural heritage. The electricity generated from these plants is being utilised for community infrastructure, powering streetlights and the village assembly hall, illustrating the successful integration of cultural preservation and sustainable energy use (Toyama City, 2021c).

Since 2003, Toyama City has pursued the establishment of a compact structure characterised by high density, multifunctional land-use development, and an integrated public transportation system. This approach encourages accessibility and convenience, together with discouraging excessive car usage and energy consumption (Ito and Kawazoe, 2022). To realise its compact city strategies, Toyama adopted a "sticks and dumplings" approach, where "sticks" represent public transportation lines and "dumplings" represent areas of high-density city functions and population. The city government encourages its citizens to live along these sticks and dumplings by providing incentives. Citizens opting to relocate within a 500-m radius of the public transportation area are offered financial incentives – up to 5,000 USD for home purchases or 100 USD for apartment rentals (personal communication, 13 September 2018). This strategy aims to increase the city's efficiency and sustainability. According to a Toyama municipal official: "we have had to make our city compact, as the city's population has been on the decline, and the declining birth rate and ageing population will further accelerate" (personal communication, 22 March 2023). The success of Toyama's compact city model, rooted in the coordinated public transportation system, led to its designation as one of Japan's Environmental Model Cities in 2009 and an Environmental Future City in 2011 (Toyama City, 2018). Internationally, the Organisation for Economic Cooperation and Development (OECD) selected the city of Toyama in 2012 as a case study for compact city strategies. In

2016, the city partnered with the World Bank in a city partnership program aimed at enhancing compact city designs (Arai *et al.*, 2020).

The compact city model significantly improves energy efficiency. Notably, investments in public transportation resulted in a reduction of 20% in petrol consumption between 2005 and 2020 (Honda, 2021). One Toyama city official commented: "Toyama's efforts to become a compact city were made to improve energy efficiency. It is all about energy efficiency" (personal communication, 22 March 2023).

Toyama's focus on and success in energy efficiency led to its designation as one of the 13 model "energy efficiency improvement cities" by the UN SE4ALL Initiative in 2014. To further accelerate energy efficiency, the city introduced the Toyama City Energy Efficiency Improvement Plan in March 2015, which aimed to double its energy efficiency from 2011 to 2030, reduce overall energy use in residences and offices through home or building energy management systems, and increase the use of renewable energy (Toyama City, 2015).

In 2018, Toyama was designated an SDGs Future City, with its programme further promoting the compact city by integrating the LRT network and implementing independent and decentralised energy management that is intended to expand sustainable city development and renewable energy use (Toyama City and IGES, 2018). The city's "Second SDGs Future City Plan for 2021–2025" involves initiatives such as reducing GHG emissions and increasing the share of renewable energy in electricity generation (Toyama City, 2021a). Furthermore, the Toyama Zero Carbon City Declaration and the Global Warming Countermeasure Promotion Plan, made public in 2021 and 2023, respectively, aim to achieve a zero-carbon city by 2050 (Toyama City, 2021b). The Toyama Zero Carbon City Declaration states that in pursuit of sustainable urban development, the city of Toyama has championed compact, public-transport-centric planning. Aligning with the national goal of zero GHG emissions by 2050, Toyama has committed to the zero-carbon city vision, intensifying environmental policies for a sustainable urban future (Toyama City, 2021b). Further, the Global Warming Countermeasures Promotion Plan outlines Toyama's goal of creating the Toyama City Global Warming Prevention Plan, stressing integration under the Global Warming Countermeasures Law (Act 117 of 1998). Renewable energy, such as solar and hydroelectric power, takes centre stage, with targets to double its adoption by 2030 and quintuple it by 2050 from the FY2021 baseline. A reduction of 46% in GHG emissions is the target by 2030, culminating in near-zero emissions by 2050. Finally, the plan underscores a comprehensive energy policy aimed at harmonising environmental sustainability with economic prosperity through unified regional efforts, fostering a zero-carbon city and broader decarbonisation (Toyama City, 2021b).

In addition to its compactness, Toyama has been striving to become a smart city by leveraging digital technology and data. In fiscal year 2022,[2] the city established the Smart City Promotion Division to align its efforts with the

SDGs and further enhance the quality of life of residents. In this context, a representative of Toyama municipal personnel commented that:

> Smart cities are highly compatible with the SDGs. Smart cities are indeed described in relation to the city's energy policies in the Second SDGs Future City Plan for 2021–2025
>
> (personal communication, 22 March 2023)

Another Toyama City official noted:

> You may wonder how the smart city concept relates to the compact city strategies. Let's put it this way: a compact city makes the city centre convenient, while a smart city covers the suburbs without disrupting the compact city strategies. Compact city is hard while smart city is soft
>
> (personal communication, 17 April 2024)

Indeed, Society 5.0 is focused on a human-centric society that seeks solutions to socio-economic challenges through a system integrating cyberspace (virtual space) and physical space (real space) (Government of Japan, 2016). That is why the integration of compact physical space and smart cyberspace is expected to be promoted within the broader context of Society 5.0 to further improve the quality of life of residents in a broader area. Specifically, this integration is anticipated to unfold across larger areas where technological applications such as sensor telecommunication network systems (low power wide area, LPWA) and lifeline platform projects can be utilised effectively (Honda, 2021). LPWA is a type of wireless telecommunication wide area network designed to allow long-range communications at a low bit rate among connected objects, such as sensors operated on a battery. The key advantages of LPWA technologies include low power usage, allowing devices to last longer with a single battery charge; wide coverage area that surpasses that of typical cellular networks; and the ability to connect a large number of devices to the network. These network systems cover 98.9% of the resident population and can assist in various tasks. This includes the operation of snow-removal equipment using GPS data logger terminals to optimise operation routes and areas, thereby promoting the energy efficiency of snow-removal-related tasks in the city (Toyama City, 2021a).

Toyama City has been addressing its energy challenges with forward-thinking practical solutions such as establishing small-scale hydroelectric power plants and utilising constant groundwater temperatures for thermal heat pumps and other applications. For example, the Sotowano Power Plant harnesses agricultural water for hydroelectric power generation and reduces CO_2 emissions (Toyama City, 2021a). With respect to energy efficiency in Toyama, groundwater is a compelling avenue for improvement. As an integral component of the natural water cycle, groundwater maintains a consistent temperature, approximately 15°C, throughout the year in the city. By capitalising on the inherent stability of the groundwater temperature, the municipal systems could enhance energy efficiency and contribute to sustainable and eco-friendly heating solutions (Toyama City, 2022b).

The findings from the analysis of Toyama highlight several key implications related to its approach to urban planning, energy efficiency, and sustainability. These findings collectively emphasise the city's innovative, sustainable, and resilient approach to urban development and energy management. Accordingly, Toyama City is a noteworthy model for other cities seeking to address energy challenges and promote sustainability through smart and compact city strategies.

3.2 Sapporo

Sapporo is the capital city of the island of Hokkaido, which is the northernmost prefecture of Japan. It is the fifth largest city in the country, with a population of two million, and serves as the economic and cultural centre for the entire prefecture. Sapporo is one of the leading smart cities in Japan (Sato *et al.*, 2022). In accordance with Society 5.0, the city is engaged in implementing a government-led smart city model project enacted by the MLIT. The city has aimed to address diverse urban challenges, particularly those related to energy, which arise because of heavy snow and the large amount of energy required for heating (Sapporo City, 2018). Household heating energy consumption in the city is approximately three times the national average (Sapporo City, 2021).

To address the energy issues, Sapporo has been developing an "Urban Energy Master Plan 2018–2050" since 2018, with the goal of becoming a zero-carbon city using renewable energy and improving energy efficiency (Sapporo City, 2018). In this Master Plan, the term "smart city" is used for a city with resilience through ICTs. In 2019, the city launched the Data-Smart City Sapporo, a website that uses data and technology to improve urban services and address city challenges. Using smart technology, the aim is to realise the *chisan chisho* of energy in the Sapporo area by creating an environment with the use of data generated in the region and held by the public and private sectors, disseminating and deploying know-how on the construction and operation of energy management infrastructure (Sapporo City, 2020). In Sapporo, district heat supply was introduced during the 1972 Winter Olympics. Since then, heat supply infrastructure has been developed with the aim of establishing a city that effectively uses energy with low environmental impact. This heat has been deployed to buildings for various purposes in an area of approximately 130 ha, the largest of such areas in Japan (Sapporo City, 2018).

One Sapporo City official noted, "we have been learning from Copenhagen, Denmark, a smart city where we have a lot in common as a city with heavy snow" (personal communication, 11 May 2023a). For example, the combination of thermoelectricity and district heat supplies with renewable energy for areal energy use, such as geothermal heat pumps and solar thermal plants, is being promoted in Denmark and other European countries (e.g. Finland and Germany). This is an effective low-carbon approach for cold regions and is also applicable to Sapporo (Sapporo City, 2018). Another Sapporo official reported that the city has implemented unique energy policies, including the use of sewage for road heating, and snow storage for cooling facilities, and is involving the city planning department in energy-related matters in pursuit of comprehensive energy conservation efforts (personal communication, 11 May 2023b).

In terms of smart technology, Sapporo City has been implementing smart grids that facilitate efficient and reliable energy distribution. These grids integrate renewable energy sources, energy storage systems, and demand–response mechanisms, contributing to the stability and resilience of a city's energy infrastructure, particularly for disaster preparedness (St-Pierre, 2023). Moreover, the city has implemented an extensive district heating and cooling (DHC) system that utilises waste heat from various sources through cogeneration. As a city representative reported, "this system efficiently provides heating and cooling to both residential and commercial buildings, leading to a significant reduction in energy consumption and subsequent GHG emissions" (personal communication, 11 May 2023c). The DHC system operates by centrally producing hot and cold water, distributing it via conduit pipes and using woody biomass (Rezaie and Rosen, 2012). This system represents an innovative approach for heating and cooling, optimising the utilisation of waste heat and renewable resources for an urban energy infrastructure that is more sustainable.

Furthermore, Sapporo utilises its snowy climate with innovative energy technologies, including snow cooling systems. As explained during the conduction of the questionnaire, these systems harness the natural coldness of snow for air conditioning and refrigeration, helping to reduce the energy demand in summer (personal communication, 11 May 2023d). A practical example of this innovation is the snow cooling system in Moerenuma Park, which was transformed from a rubbish-disposal site. Each year around mid-March, approximately 1,700 tons of slightly melted snow is collected and transported to a snow storage facility through snow transfer tubes (Moerenuma Park, 2023). The stored snow gradually melts during the summer months, with the cold water generated from this process circulating through buildings. Furthermore, the cold air produced is used to cool servers in data centres, reducing cooling costs from 1.5 billion JPY to 300 million JPY[3] (Sapporo City, 2018). When city structures are renovated, snow storage facilities must be installed and used for cooling to save energy (Sapporo City, 2018). Similar to Toyama City, Sapporo employs thermal energy from sewage, which maintains a consistent temperature of approximately 15°C throughout the year. This thermal energy is utilised to melt snow as it flows through the city's sewage system (personal communication, 11 May 2023d).

The smart city and energy initiatives of Sapporo offer a promising model for sustainable urban development, providing inspiration and lessons for other cities with a snowy climate. This innovative approach could contribute to environmental sustainability, energy efficiency, and improved quality of life, showing a transformative blueprint for urban centres worldwide.

3.3 Kumamoto

Kumamoto is the capital city of Kumamoto Prefecture, located on Kyushu Island in southeastern Japan. As of June 2023, its population was 733,000 residents (Kumamoto City, 2023a). The city is situated in a mountainous basin and is undergoing a shift towards new urbanisation characterised by denser transport networks. This transformation is reshaping the urban spatial distribution of the population, employment, land use and travel patterns, which, ultimately, led to a noticeable reduction in energy consumption (Yin *et al.*, 2013).

The commitment of the city to global environmental issues and sustainable urban development has been solidified through its recent sister city agreement with Heidelberg, Germany, known for its commitment to climate neutrality. The signing ceremonies took place first in Heidelberg in May 2021 and later in Kumamoto in September of the same year, accompanied by events focusing on global environmental issues, and leading to the development of programmes across a broader range of fields. This partnership, based on the principles of "common responsibility for peace and the environment", underscores the dedication of both cities to climate and paves the way for joint initiatives that address critical global challenges (Kumamoto City, 2024).

In collaboration with industry and academia, Kumamoto City has been pursuing the vision of becoming a smart energy city by adopting a comprehensive approach that extends beyond the mere utilisation of new technologies (Kumamoto City, 2023b). One of the key drivers of smart energy initiatives is the need to address regional challenges effectively, such as the increasing frequency and severity of natural disasters (e.g. typhoons or earthquakes), chronic traffic congestion and delays in digitising government administration and city functions (Kumamoto City, 2023b).

When major disasters strike, reliance on automobiles raises concerns regarding the effectiveness of emergency response efforts (Liu *et al.*, 2020). To tackle this challenge, as done by Toyama and Sapporo, since 2014, Kumamoto has been working towards creating a well-connected, integrated and compact city network characterised by multiple central urban hubs and regional bases that are all linked through public transportation systems to foster collaborative coordination among the different city components (Kumamoto City, 2017). The initiatives also focus on expanding the use of renewable energy in public facilities by developing an independent and decentralised energy system through government-subsidised projects and partnerships with surrounding municipalities and local companies (Kumamoto City, 2023b).

According to respondents from the Kumamoto municipal government, the city uses its status as a *seirei shitei toshi* ("government-designated city")[4] under *chihō jichi hō* ("local autonomy law").[5] These cities have been designated by government ordinance to have a population greater than 500,000 and are granted a level of autonomy similar to that of a prefecture. With this status, Kumamoto City demonstrates leadership in collaborating with neighbouring municipalities to work towards achieving the SDGs (personal communication, 19 May 2023). A government-designated city is a classification of the administrative divisions in Japan, specifically granted to larger cities within local governments. As these cities have the same level of autonomy as prefectures, they are endowed with authority and responsibility for significant administrative and urban planning tasks.

In March 2021, Kumamoto formulated a Global Warming Prevention Action Plan for the Kumamoto Coordinated Core City Region (involving 18 neighbouring municipalities). The aim of the strategy is to achieve zero-carbon emissions by 2050. As one Kumamoto official explained:

> Decarbonisation cannot be solved by a single municipality. Therefore, a macroscopic perspective was required. Amid concerns about the increase and intensification of natural disasters and biodiversity crises due to climate change, it is necessary for the city and surrounding areas to realise a sustainable decarbonised circulatory symbiosis zone by fully utilising and circulating renewable energy within the zone in order to make the local economy and the lives of citizens sustainable
> (personal communication, 19 May 2023a)

A notable feature of a city's energy policy is the creation of a self-sustaining and decentralised energy system, in which even if one facility experiences a power outage, others can cover it. For instance, a Kumamoto municipal official indicated:

> In case Kyushu Electric Power Company experiences a power outage, to secure electricity power source, the city has established an agreement with Nissan Motor Corporation to use EVs in the event of a disaster
> (personal communication, 19 May 2023b)

A challenge in a renewable energy system is uncertainty regarding the amount of electricity that could be generated. A Kumamoto city representative reported:

> If the supply of renewable energy is not sufficient, we have to buy energy from Kyushu Electric Power Company, but it is expensive. As with any municipality, the cost aspect must be taken into consideration
> (personal communication, 19 May 2023c)

The official noted:

> In recent years, large-scale natural disasters have occurred in many regions. Efforts towards disaster prevention and mitigation are indispensable for cities in Japan and abroad. In particular, the maintenance of lifelines should be given top priority. We believe Kumamoto City's experiences and lessons learned from the 2016 Kumamoto earthquakes can contribute to disaster response in other regions, while our independent

> and decentralised energy initiatives can make a significant contribution to the disaster prevention and disaster mitigation efforts of cities in Japan and abroad
>
> (personal communication, 19 May 2023c).

A notable initiative involving electricity generation from waste is the municipal Smart Energy Kumamoto, a community-based energy company established in 2018. The establishment of this company was driven by several factors, including the promotion of the 2016 Kumamoto City Earthquake Reconstruction Plan, efforts to create low or zero-carbon cities in accordance with international agreements such as the Paris Agreement (adopted in 2015) and SDGs, and the need to reduce electricity costs. The company provides 30% of the electricity used by city facilities, resulting in a reduction in electricity charges of 160 million JPY.[6] A portion of these funds can be reinvested in subsidising citizens and businesses to adopt energy-saving equipment, solar power generation or storage batteries, as well as building zero-energy homes. Using large storage batteries has contributed to reducing peak electricity demand and enhancing disaster prevention efforts (personal communication, 19 May 2023d). Subsequently, Kumamoto City has invested and became actively involved in managing the company. As Deguchi (2018) notes, smart energy community projects are anchored within government strategies such as the Kyoto Protocol Target Achievement Plan, which the Cabinet formulated in April 2005 (revised in March 2008). Therefore, the Smart Energy Kumamoto initiative has been subsidised by the national government.

According to a Kumamoto official:

> It is essential to acknowledge that the paths towards decarbonisation and SDGs vary from one region to another. In cases where the existing systems or policies are insufficient, the creation of new entities such as Smart Energy Kumamoto is necessary
>
> (personal communication, 19 May 2023d)

Another Kumamoto municipal representative concluded:

> Through capitalising on its inherent strengths and fostering collaboration with key stakeholders, Kumamoto City envisions itself evolving into a trailblazing smart energy model city, setting a benchmark for other regions to follow
>
> (personal communication, 19 May 2023e)

Kumamoto's initiatives highlight a region-specific approach to decarbonisation and sustainable development with an emphasis on innovation and collaboration, as exemplified by Smart Energy Kumamoto and the ambition to become a pioneering smart energy municipality.

4. Discussion

The goals of these three cities align with common aspirations to achieve smart, resilient, and environmentally responsible urban development. However, the strategies and initiatives of the cities exhibit remarkable diversity, reflecting the unique challenges and opportunities presented by their respective geographic, climatic, and socioeconomic contexts. In this light, this section offers a comprehensive analysis of the commonalities and differences among the cities of Toyama, Sapporo, and Kumamoto regarding their approaches to urban planning, energy efficiency, and sustainability.

4.1 Commonalities Among Toyama, Sapporo, and Kumamoto as SDGs Future Cities

In the pursuit of sustainable urban development, Toyama, Sapporo, and Kumamoto emerge as trailblazers, each manifesting a fervent commitment to the principles of sustainability. These SDGs Future Cities demonstrate their resolve by actively reducing GHG emissions, embracing renewable energy, and fostering urban designs that prioritise efficiency and reduce reliance on fossil fuels. Their collective efforts in innovation and technology, urban compactness, and collaborative engagement exemplify a multifaceted approach to sustainable development. The following sections explore these common threads, delving into how each city uniquely navigates the challenges and opportunities presented by their environment, society, and technological landscapes – contributing valuable insights into the global dialogue on smart, resilient, and sustainable cities.

4.1.1 Commitment to Sustainability

The three cities share a strong commitment to sustainability (Toyama City, 2018; Sapporo City, 2018; Kumamoto City, 2023b), all recognising the importance of addressing environmental challenges, reducing GHG emissions, and promoting sustainable development to improve the quality of life of their residents and ensure a resilient future. They are also actively investing in renewable energy sources to reduce their reliance on fossil fuels and promote the use of cleaner energy alternatives, such as renewable energy sources including geothermal, solar, and woody biomass-derived bioenergy (Toyama City and IGES, 2018; Sapporo City, 2018; Kumamoto City, 2023b). These commitments to renewable energy align with global efforts to combat climate change.

4.1.2 Urban Compactness

All three cities place strong emphasis on urban compactness and high-density development (Ito and Kawazoe, 2022; Sapporo City, 2018; Kumamoto City, 2017). These cities promote the use of public transportation while discouraging excessive reliance on automobiles, thereby contributing to reducing energy consumption within their respective regions. This concerted effort towards sustainable urban design aligns with their shared commitment to a greener and future that is more energy efficient.

4.1.3 Innovation and Technology

The three cities leverage innovation and technology to achieve their sustainability goals by incorporating smart city concepts, digital technology and data-driven solutions to improve energy efficiency (personal communication, 22 March 2023; Sapporo City, 2020), disaster preparedness (St-Pierre, 2023; personal communication, 19 May 2023c) and overall urban quality (Toyama City and IGES, 2018; Sapporo City, 2018; Kumamoto City, 2023b). The cities explore and implement new technologies to optimise energy use and enhance the well-being of their residents.

4.1.4 Collaboration and Engagement

Collaboration among various stakeholders, including government bodies, industry partners, academic institutions, and communities is a shared feature in all three cities. Toyama partners with local companies to develop independent and decentralised energy systems (Toyama City, 2015). Sapporo discussed its plan of using electric vehicles in the event of a natural disaster with private companies such as Nissan Motor Corporation, demonstrating its collaboration with industry partners (Sapporo City, 2018).

Furthermore, Toyama has partnered with the World Bank in a city partnership programme aimed at enhancing compact city designs, and the city has received support and recognition from organisations such as the OECD (Arai *et al.*, 2020). *Sapporo learned from Copenhagen how to address its* unique challenges, especially those related to heavy snowfall (Sapporo City, 2018; personal communication, 11 May 2023). Kumamoto collaborates with neighbouring municipalities towards achieving its SDGs, leveraging its status as a government-designated city (Kumamoto City, 2023b; personal communication, 19 May 2023a). The cities recognise that addressing complex urban challenges such as demographic shifts (Toyama City, 2023), heavy snow and energy costs (Sapporo City, 2021), and natural disasters (Kumamoto City, 2023b) require a multi-stakeholder approach, and they actively engage with stakeholders to develop and implement their sustainability initiatives.

4.2 Differences Among Toyama, Sapporo, and Kumamoto as SDGs Future Cities

Distinct differences mark the sustainable paths of Toyama, Sapporo, and Kumamoto, each navigating unique energy landscapes. These cities, while unified in striving towards sustainability and innovation, customise their strategies to local needs and global sustainability standards. Their tailored initiatives, partnerships, and governance models not only enhance their own urban landscapes but also offer diverse insights into building adaptable, resilient, and sustainable smart cities worldwide. The sections that follow delve into the specifics of their journeys, discussing how each city interprets and applies the principles of sustainability to their unique urban challenges and opportunities.

4.2.1 Climate and Energy Challenges

Although all three cities are confronted by energy-related challenges, the nature of these challenges varies. The city of Toyama emphasises maximising the use of local energy sources, such as hydroelectric power and geothermal energy to meet its energy demands and enhance its resilience against natural disasters (Toyama Prefecture, 2018). The city of Sapporo focuses on addressing its high energy consumption for heating owing to its cold climate, leading to innovative solutions such as snow cooling systems (Sapporo City, 2018). The city of Kumamoto has to deal with natural disasters, driving its efforts to create a self-sustaining and decentralised energy system to ensure reliability during adverse events (Kumamoto City, 2023b).

4.2.2 Specific Initiatives

All three Japanese cities share common sustainability goals such as GHG reduction and renewable energy adoption; however, their specific initiatives set them apart. The compact city model of Toyama with a focus on energy efficiency has garnered international recognition and partnerships with recognised organisations such as the World Bank (Arai *et al.*, 2020). In Sapporo, the emphasis on snow cooling systems addresses its unique climatic challenges. The status of Kumamoto as a government-designated city allows it to collaborate with neighbouring municipalities and lead collaborative efforts to achieve zero-carbon emissions by 2050 (personal communication, 19 May 2023a). Sapporo collaborates with other smart cities globally, such as Copenhagen, to learn from their experiences and adopt best practices (personal communication, 11 May 2023a). Toyama City, in partnership with Regency Tabanan, has installed four micro-hydroelectric plants in Jatiluwih, powering streetlights and an assembly hall. Kumamoto has strengthened its commitment to addressing environmental challenges by forming a sister city pact with Heidelberg, a city also focused on achieving climate neutrality.

4.3 Implications

The implications drawn from the initiatives of Toyama, Sapporo, and Kumamoto extend beyond their regional impacts to offer a blueprint for global urban development. These cities exemplify the integration of smart technologies, collaborative governance, and climate-responsive strategies, providing a path for cities worldwide to navigate the complexities of sustainable development and energy management. This section carefully considers their collective experiences and the lessons they offer towards fostering resilient, adaptive, and technologically advanced urban environments at a global scale.

4.3.1 Global Relevance

As rapid urbanisation continues in cities worldwide, the strategies implemented by Toyama, Sapporo, and Kumamoto are gaining global relevance because they confront common urban challenges, which are especially relevant to Asian cities. These challenges encompass urban growth, sustainability, energy efficiency, and resilience against disasters. Their approaches in crafting sustainable and resilient urban environments offer practical blueprints for cities aiming to forge a sustainable future.

4.3.2 Collaborative Governance

The collaborative governance models employed in the analyses of the three cities highlight the importance of involving stakeholders in sustainability initiatives. Government bodies, industry partners, academic institutions, and communities play a crucial role in driving such efforts. This approach fosters a sense of shared responsibility and accountability for achieving sustainability goals.

4.3.3 Climate Adaptation

The endeavours of Toyama to build a compact city, response of Sapporo to its cold climate, and efforts of Kumamoto to address natural disasters by establishing Smart Energy Kumamoto demonstrate the importance of tailoring sustainability initiatives to local conditions. These examples prove how innovative solutions can be developed to adapt to specific challenges and ensure the resilience of urban areas confronted by environmental and climate threats.

4.3.4 Energy Transition

The resolute commitment of Toyama, Sapporo, and Kumamoto to utilising renewable energy sources is not only a local endeavour but also a significant stride in agreement with global efforts to shift away from environmentally harmful fossil fuels. The forward-thinking initiatives championed by these cities play a pivotal role in curbing GHG emissions, ultimately contributing to a reduction in the overall carbon footprint.

For Japan, it is most important to adhere to the trend of implementing initiatives to reduce GHG emissions, as the country houses dense urban populations. Moreover, committing to curbing emissions through innovative city-level strategies could significantly reduce the carbon footprint of Japan. Such initiatives set a sustainable precedent for other nations to follow, reinforcing Japan's leadership role in combating climate change.

4.3.5 Smart Cities

The integration of smart city concepts and advanced digital technologies within these three urban centres does not stop at boosting energy efficiency; it also ushers in a transformative shift in the overall quality of life experienced by their residents. By harnessing data-driven solutions and technology-infused infrastructures, these Japanese cities are vital elements of a global evolution of smart and connected urban landscapes. Their approach serves as an inspirational model at different levels, namely national (in Japan), regional (in Asia), and global. Accordingly, the city initiatives demonstrate how interwoven technology and urban planning could lead to enhanced living standards, increased sustainability, and a prosperous and interconnected future for cities worldwide.

5. Conclusion

The experiences of Toyama, Sapporo, and Kumamoto encapsulate the forward momentum towards smart and sustainable urban development. These cities, each in their unique way, demonstrate a steadfast dedication to sustainability, leveraging their distinct local resources and governance structures to craft innovative solutions to climate challenges. Their paths offer insightful paradigms of urban energy efficiency and development, underscoring the critical role of adaptability and collaboration in fostering smart and resilient urban environments. As these cities continue to refine their strategies, they not only contribute to the green transformation of Japan but also inspire other global urban centres to embark on their journeys towards a sustainable and prosperous future.

Acknowledgements

I wish to express my sincere gratitude to Dr Nobutaka Ito, Professor Emeritus, Faculty of Bioresources, Mie University, for his guidance and sharing his expertise in environmental and energy research. I also extend my appreciation to distinguished personnel representing the city governments of Toyama, Sapporo, and Kumamoto.

Notes

1 In Japanese: 地産地消 [*chisanchishō*].
2 Japanese fiscal year 2022 covers the period from April 2022 to March 2023.
3 From approximately 10 million USD to approximately 2 million USD.

4 In Japanese: 政令指定都市 [*seirei shitei toshi*].
5 In Japanese: 地方自治法 [*chihōjijihō*].
6 Approximately 1.1 million USD.

References

Agency for Natural Resources and Energy (2014) "Fourth Energy Plan". In Japanese: 第4次エネルギー基本計画 [*Dai 4-ji Enerugī Kihon Keikaku*]. Available at: https://www.enecho.meti.go.jp/category/others/basic_plan/pdf/140411.pdf (Accessed: 30 April 2024).

Arai, Y., Levine, D., Miki-Imoto, H. and Mitsuhiro, Y. (2020) *The Development Story of Toyama*. Washington, DC: World Bank.

Balaban, O. and Puppim de Oliveira, J.A. (2022) "Finding Sustainable Mobility Solutions for Shrinking Cities: The Case of Toyama and Kanazawa", *Journal of Place Management and Development*, 15(1), pp. 20–39.

Battarra, R., Fistola, R. and La Rocca, R.A. (2016) "City SmartNESS: The Energy Dimension of the Urban System", in R. Papa and R. Fistola (eds) *Smart Energy in the Smart City: Urban Planning for a Sustainable Future*. Cham: Springer, pp. 1–23.

Bowen, G.A. (2009) "Document Analysis as a Qualitative Research Method", *Qualitative Research Journal*, 9(2), pp. 2–40.

Cabinet of Japan (2018) "Presentation Ceremony for Certificate of the FY2018 'SDGs Future Cities". Available at: https://japan.kantei.go.jp/98_abe/actions/201806/_00040.html (Accessed: 30 April 2024).

Cardno, C. (2018) "Policy Document Analysis: A Practical Educational Leadership Tool and a Qualitative Research Method", *Educational Administration: Theory and Practice*, 24(4), pp. 623–640.

Deguchi, A. (2018) "From Smart City to Society 5.0", in Hitachi-UTokyo Laboratory (ed) *Society 5.0: A People-Centric Super-Smart Society*. Singapore: Springer Singapore, pp. 43–66.

DeWit, A. (2013) "Japan's Rollout of Smart Cities: What Roles for the Citizens?", *The Asia-Pacific Journal*, 11(24), pp. 1–11.

Government of Japan (2016) "The Fifth Science and Technology Basic Plan". Available at: https://5x5.wirelesswatch.jp/docs/S5-plan.pdf (Accessed: 30 April 2024).

Hollands, R. (2008) "Will the Real Smart City Please Stand Up?" *City*, 12(3), pp. 303–320.

Honda, S. (2021) "A Historical Study of Toyama City's Urban Development and Leadership". In Japanese: 富山市のまちづくりとリーダーシップの歴史的考察 [*Toyama-shi no Machizukuri tori-da- shippu no Rekishiteki Kōsatsu*], *Trends in the Science*, 26(6), pp. 36–41.

Ito, H. and Kawazoe, N. (2022) "Promoting Transportation Policies in the Context of Compact City Strategies: The Case of Toyama City, Japan", *The Annals of Regional Science*, 71, pp. 775–797.

Kumamoto City (2017) "The Revised Second Master Plan". In Japanese: 第2次熊本市都市マスタープラン [*Dai 2-ji Kumamoto-shi Toshi Masutā Puran*]. Available at: https://www.city.kumamoto.jp/common/UploadFileDsp.aspx?c_id=5&id=14791&sub_id=3&flid=114841 (Accessed: 30 April 2024).

Kumamoto City (2023a) "Kumamoto City's Estimated Population". In Japanese: 推定人口[*Suitei Jinkō*]. Available at: https://www.city.kumamoto.jp/hpkiji/pub/detail.aspx?c_id=5&id=2382 (Accessed: 30 April 2024).

Kumamoto City (2023b) "Smart City Kumamoto Promotion Strategies". In Japanese: スマートシティくまもと推進戦略 [*Sumāto Shiti Kumamoto Suishin Senryaku*]. Available at: https://www.city.kumamoto.jp/common/UploadFileDsp.aspx?c_id=5&id=38069&sub_id=27&flid=292657 (Accessed: 30 April 2024).

Kumamoto City (2024) "Sister City: Heidelberg". In Japanese: 友好都市・ハイデルベルク市 [*Yūkō Toshi Haideruberuku-shi*]. Available at: https://www.city.kumamoto.jp/hpKiji/pub/detail.aspx?c_id=5&id=3410 (Accessed: 30 April 2024).

Library of Congress (2020) "Japan: 'Super City' Law Enacted". Available at: https://www.loc.gov/item/global-legal-monitor/2020-08-18/japan-super-city-law-enacted/ (Accessed: 30 April 2024).

Liu, H., Homma, R. and Iki, K. (2020) "Case Study of the 2016 Kumamoto Earthquake: The Disaster Response Capability of Kumamoto Compact City, Japan", *MATEC Web of Conferences*, 331, pp. 1–10.

Masuda, H., Kawakubo, S., Okitasari, M. and Morita, K. (2022) "Exploring the Role of Local Governments as Intermediaries to Facilitate Partnerships for the Sustainable Development Goals", *Sustainable Cities and Society*, 82, pp. 1–12.

Matsumoto, T., Esaka, T. and Izumiyama, R. (2023). "A Study on Initiatives Related to Regional Characteristics of Smart Cities". In Japanese: スマートシティの地域特性における取り組みに関する研究 [*Sumāto Shiti no Chiiki Tokusei ni Okeru Torikumi ni Kansuru Kenkyū*], *Reports of the City Planning Institute of Japan*, 21, pp. 351–356.

Moerenuma Park (2023) "Environmental Initiatives". Available at: www.moerenumapark.jp/environment (Accessed: 30 April 2024).

Morgan, H. (2021) "Conducting a Qualitative Document Analysis", *The Qualitative Report*, 27(1), pp. 64–77.

Nago, M., Murata, H. and Ito, H. (2021) "Recent Trends of Smart City and Its Infrastructure Toward SDGs: From Case Analysis of Municipalities by Industrial Organization". In Japanese: SDGs志向スマートシティ・インフラ [*SDGs Shikō Sumāto Shiti Infura*], *Development Engineering*, 41(1), pp. 61–64.

Nakamura, K. and Krizaj, D. (2020) "A Service Platform Model Through Building Smart/Super Cities: Potentials for Sustainable Community OS From City-Building OS Model", *Development Engineering*, 40(1), pp. 127–133.

Papadopoulou, C.-A. (2021) "Technology and SDGs in Smart Cities Context", in A. Visvizi and R.P. del Hoyo (eds) *Smart Cities and the UN SDGs*. Amsterdam: Elsevier, pp. 45–58.

Quitzow, L. and Rohde, F. (2022) "Imagining the Smart City Through Smart Grids? Urban Energy Futures Between Technological Experimentation and the Imagined Low-Carbon City", *Urban Studies*, 59(2), pp. 341–359.

Rezaie, B. and Rosen, M.A. (2012) "District Heating and Cooling: Review of Technology and Potential Enhancements", *Applied Energy*, 93, pp. 2–10.

Sapporo City (2018) "Urban Energy Master Plan 2018–2050". In Japanese: 都心エネルギーマスタープラン [*Toshin Enerugī Masutā Puran*]. Available at: https://www.city.sapporo.jp/kikaku/downtown/toshin-energy/documents/mp_1.pdf (Accessed: 6 January 2025).

Sapporo City (2020) "ICT Utilisation Strategy". In Japanese: 札幌市ICT活用戦略2020 [*Sapporo-shi ICT Katsuyō Senryaku 2020*]. Available at: https://www.city.sapporo.jp/kikaku/ictplan/documents/gaiyouban.pdf (Accessed: 30 April 2024).

Sapporo City (2021) "Sapporo City SDGs Future City Plan". In Japanese: 札幌市SDGs未来都市計画 [*Sapporo-shi SDGs Mirai Toshi Keikaku*]. Available at: https://www.city.sapporo.jp/kankyo/sdgs/documents/sdgs_plan2.pdf (Accessed: 30 April 2024).

Sato, T., Muraki, M. and Sunaga, D. (2022) "Study on Efficient Administrative Management of Local Governments Through Interdisciplinary Cooperation on Smart City Project: A Case Study in Sapporo City". In Japanese: スマートシティ事業を通じた自治体の分野間連携による効率的な行政運営のあり方に関する考察 [*Sumāto Shiti Jigyō o Tsūjita Jichitai no Bunyakan Renkei ni yoru Kōriteki na Gyōsei Unei no Arikata ni Kansuru Kōsatsu*], *AIJ Journal of Technology and Design*, 28(70), pp. 1420–1425.

Singh, A. and Singla, A.R. (2020) "Constructing Definition of Smart Cities from Systems Thinking View", *Kybernetes*, 50(6), pp. 1919–1960.
Söderström, O., Paasche, T. and Klauser, F. (2014) "Smart Cities as Corporate Storytelling", *City*, 18(3), pp. 307–320.
Sokołowski, M.M. (2021) "Models of Energy Communities in Japan (Enekomi): Regulatory Solutions From the European Union (Rescoms and Citencoms)", *European Energy and Environmental Law Review*, 30(4), pp. 149–159.
Sokołowski, M.M. (2022) *Energy Transition of the Electricity Sectors in the European Union and Japan*. Cham: Palgrave Macmillan.
Somayya, M. and Ramaswamy, R. (2016) "Amsterdam Smart City: Fishing Village to Sustainable City", *WIT Transactions on Ecology and the Environment*, 204, pp. 831–842.
Stewart, J. (2012) "Multiple-Case Study Methods in Governance-Related Research", *Public Management Review*, 14(1), pp. 67–82.
St-Pierre, E. (2023) "Contested Energy Futures in Hokkaido: Speculating With European Renewable Energy Models", in S. Sareen and K. Muller (eds) *Digitisation and Low-Carbon Energy Transitions*. Cham: Palgrave Macmillan, pp. 53–72.
Su, Y. and Fan, D. (2023) "Smart Cities and Sustainable Development", *Regional Studies*, 57(4), pp. 722–738.
Thornbush, M. and Golubchikov, O. (2021) "Smart Energy Cities: The Evolution of the Energy-City-Sustainability Nexus", *Environmental Development*, 39(5), pp. 1–11.
Toyama City (2015) "Improving Energy Efficiency". Available at: https://www.city.toyama.lg.jp/_res/projects/default_project/_page_/001/005/126/gaiyou.eng.pdf (Accessed: 30 April 2024).
Toyama City (2018) "Toyama City the Sustainable Development Report". Available at: https://www.local2030.org/pdf/vlr/english-vlr-toyama-city-japan-2018.pdf (Accessed: 30 April 2024).
Toyama City (2021a) "The Second Toyama City SDGs Future City Plan". In Japanese: 第2次富山市SDGs未来都市計画 [*Dai 2-ji Toyama-shi SDGs Mirai Toshi Keikaku*]. Available at: https://sdgs.city.toyama.lg.jp/common/pdf/sdgsmiraitoshikeikaku.pdf (Accessed: 30 April 2024).
Toyama City (2021b) "Toyama City Zero Carbon City Declaration". In Japanese: 富山市ゼロカーボンシティ [*Toyama-shi Zero Kābon Shiti*]. Available at: https://www.city.toyama.lg.jp/_res/projects/default_project/_page_/001/005/150/01zerocarbonsengensyo.pdf (Accessed: 30 April 2024).
Toyama City (2021c) "International Cooperation Future City Toyama". Available at: https://www.city.toyama.lg.jp/_res/projects/default_project/_page_/001/005/134/brochure_2021.pdf (Accessed: 8 February 2024).
Toyama City (2022a) "Toyama City's Smart City Promotion Vision: Mid-Term Report". In Japanese: 富山市スマートシティ推進ビジョン中間報告 [*Toyama-shi Sumāto Shiti Suishin Bijon Chūkan Hōkoku*]. Available at: https://www.city.toyama.lg.jp/_res/projects/default_project/_page_/001/003/052/visonchukan220328.pdf (Accessed: 18 February 2024).
Toyama City (2022b) "Effective Use of Groundwater". In Japanese: 地下水の適正利用 [*Chikasui no Tekisei Riyō*]. Available at: https://www.city.toyama.lg.jp/kurashi/gomi/1010244/1010245/1005246.html (Accessed: 30 April 2024).
Toyama City (2024) "Statistics of Toyama City in 2024". In Japanese: 富山市の人口 [*Toyama-shi no jinkō*]. Available at: https://www.city.toyama.lg.jp/shisei/1001818/1010951/1001823/1011659.html (Accessed: 5 January 2025).
Toyama City and IGES (2018) *Toyama City Sustainable Development Goals Report: Compact City Planning Based on Polycentric Transport Networks*. Hayama: IGES.
Toyama Prefecture (2022) "Toyama Prefecture Renewable Energy Vision". In Japanese: 富山県再生可能エネルギービジョン [*Toyama-ken Saisei Kanō Enerugī Bijon*].

Available at: https://www.pref.toyama.jp/documents/25189/mat01.pdf (Accessed: 6 January 2025).
Yin, R. (2018) *Case Study Research and Applications: Design and Methods*. Thousand Oaks Trees, CA: SAGE.
Yin, Y., Mizokami, S. and Maruyama, T. (2013) "An Analysis of the Influence of Urban Form on Energy Consumption by Individual Consumption Behaviors From a Microeconomic Viewpoint", *Energy Policy*, 61, pp. 909–919.
Zappa, M. (2020) "Smart Energy for the World: The Rise of a Technonationalist Discourse in Japan in the Late 2000s", *International Quarterly for Asian Studies*, 51(1–2), pp. 193–222.
Zuccala, M. and Verga, E.S. (2017) "Enabling Energy Smart Cities Through Urban Sharing Ecosystems", *Energy Procedia*, 111, pp. 826–835.

11 Smart City Development Underway

Lessons from Shin-Sapporo's Smart City

Carin Holroyd

1. Introduction

Over the last two decades, interest in how to develop 21st-century cities has skyrocketed. The United Nations predicts that by 2050, 68% of the world's population will live in urban areas (United Nations Department of Economic and Social Affairs, n.d.). Cities collectively are, and will increasingly be, therefore, responsible for the consumption of most of the world's energy and the production of most of its waste and greenhouse gas emissions; developing ways to make cities more environmentally sound is of vital importance. Cities have been working on how to best use technology to improve the quality, performance, and cost of various kinds of urban architecture. As information and communications technologies (ICT) improved and access to data increased, more thought was given to how this could best be used to deal with urban problems, particularly those associated with the environment. National and municipal governments and corporations set up pilot projects (called smart cities, future, eco or low-carbon cities) to test out the feasibility of using ICT to improve energy and transportation infrastructure and create more environmentally friendly, safe, and enjoyable homes and communities.

A 2014 report on "Smart Cities in Japan" identified four main goals for smart city projects: fostering energy security and efficiency (by lowering energy consumption and helping introduce renewables); security/survival in times of disaster; boosting economic development by showcasing large Japanese companies; and, as smart cities are often constructed in underutilised areas of the city, revitalising the area and the local economy (Pham, 2014). The focus of this chapter, the Shin-Sapporo Smart City, was designed with all of these objectives in mind. Before describing the Shin-Sapporo Smart City and the lessons that can potentially be learned from this new smart city project, some background on smart cities in Japan is essential.

2. Japanese Smart Cities

Japan was one of the first countries involved in smart city construction (Su *et al.*, 2022). The earliest discussions in the Japanese government about the

DOI: 10.4324/9781003471448-11

potential of an environmentally friendly city were in a 1989 Environmental White Paper (Ministry of the Environment, 1989). A few years later the then Ministry of Construction launched an environmentally friendly model city project (Okubo *et al.*, 2022). The smart city movement emerged out of a long-standing interest in green growth (Holroyd, 2018). A national programme in support of municipal efforts to decrease greenhouse gas emissions – the Eco Model Cities project – was launched in 2008. Thirteen cities were selected for funding based on their proposals for achieving major reductions in greenhouse gas emissions. Support for low-carbon cities continued, especially after the Great East Japan Earthquake in 2011. The Eco-Model City programme was integrated into a Future Cities Initiative that encouraged cities to deal with both environmental challenges and those of an ageing society (Future City, n.d.). In 2020, Japan enacted the Super City Law (the amended National Strategic Special Zone Act)[1] to digitally transform cities through the use of artificial intelligence (AI) and big data and private–public sector collaboration (Hirayama, 2021).

The Ministry of Economy, Trade and Industry (METI) called for Smart City proposals in January 2010. The private sector (primarily real estate and construction companies) played a leading role in the development of these proposals. METI selected Kitakyushu, Yokohama, Kyoto's Keihanna District (Kansai Science City), and Toyota City to carry out large-scale Smart City demonstration projects from 2010 to 2015. Each city had CO_2 emissions reduction goals and different plans related to energy (solar energy, Electric Vehicle (EV) car sharing, smart homes, etc.).

The Japanese private sector has always been active with respect to smart community projects. These pilots have been designed to demonstrate advanced technology that the companies hope will attract both domestic and foreign customers (this includes companies that make key components of smart cities like Sekisui House, a pioneer in the construction of zero-emissions homes). A study of Japanese smart communities by Okubo *et al.* (2022) divided them into three overlapping priority areas: information technology (including energy management, energy use visualisation), town planning (including disaster management and low carbon use), and environment and energy (renewable energy use and energy efficiency). So, although energy efficiency and low carbon emissions are not the only focus of all Japanese smart cities, they are usually two of the main objectives.

There are numerous examples of Japanese private sector smart cities of varying sizes and characteristics. Toshiba was involved in smart community projects in Yokohama and Ibaraki in the early 2010s. Panasonic has been very involved with smart city developments. In 2012, Panasonic's subsidiary PanaHome launched Smart City Shioashiya "Solar-Shima", 400 homes and an 83-unit apartment complex focused on using solar energy and reducing energy consumption. Panasonic has created three Sustainable Smart Towns (SST). Phase one of the first of these, Fujisawa SST opened in 2014 on one of the company's abandoned factory sites in a suburb of Fujisawa City, 50 km.

outside of Tokyo. Fujisawa SST is now home to 600 households. One of the main priorities of this SST is the efficient use of energy. Homes are equipped with solar panels, home energy management systems, and storage batteries. The president of Fujisawa SST believes that, for smart cities, the future "lies in their ability to address global warming by being carbon neutral, and in their ability to improve people's well-being" (Hornyak, 2022).

Panasonic's second SST is Tsunashima (located in Yokohama, Kanagawa Prefecture) which has a more commercial focus than a residential one. The town's electricity and heating are supplied by a gas cogeneration system similar to the one in Shin-Sapporo, the smart city project to be discussed in this chapter. Hydrogen fuel cells and vehicle refuelling are available and various energy management systems are being tested (Panasonic, 2022). Suita SST opened in 2022. Its main focus is addressing Japan's ageing population by creating a multi-generational community that uses technology to make the older adults safer. However, it also aims to use 100% renewable energy and be energy self-sufficient in the event of a disaster (Hornyak, 2022; Wray, 2022).

Apart from Panasonic, other Japanese companies are also active in the field of smart cities. For instance, Daiwa House (a major housing construction company) created SMA x ECO Town Harumidai, a community of 65 homes powered by solar energy systems and storage batteries in all homes and common areas. Energy usage visualisation systems are available in each home and there is an electric car–sharing programme. Toyota announced plans for a city of the future at the base of Mt. Fuji. Named Woven City, it will be both a "living laboratory" for new technologies and home to 2,000 people who will live in a hydrogen-powered (Toyota is releasing a portable hydrogen fuel cell) fully sustainable ecosystem (Regan, 2022; Robertson, 2022; Toyota, 2023). Moreover, municipalities have also launched smart city initiatives. Bonjono Kitakyushu Smart City Project, M-Smart City Kumagaya, and Ayameike Housing Development (a public–private partnership) would be examples (Gondokusuma, Kitagawa, and Shimoda, 2019). There has also been a substantial growth in urban and rural energy communities across Japan (see Sokołowski, 2021).

3. Shin-Sapporo's Smart City Project

Sapporo, the fifth largest city in Japan and the capital of Hokkaido, has been committed to being an environmentally friendly city for the last two decades. In 2008, the Sapporo Sustainability Declaration – a commitment by universities to contribute towards sustainability – was adopted by the Group of Eight (G8) University Summit at a meeting held that year in Sapporo (Hokkaido University, n.d.). Since then, Sapporo has been noted for its efforts to be a low-carbon and resilient city. Quickly, it began to adopt renewable sources of energy like solar, biomass, and biodiesel. Sapporo City's Developmental Strategic Vision published in 2013 set forth a ten-year vision for the city that described becoming a low-carbon society as one of its three pillars (City of Sapporo, 2013).

The Strategic Vision report showed that in Sapporo the biggest sources of carbon dioxide emissions were civilian households and the civilian business sector (offices, hotels, and other service businesses), much more so than the industry or transportation sectors as in Japan overall (City of Sapporo, 2013). Therefore, if Sapporo wanted to reduce its CO_2 emissions, it would need to begin with its residents. That is why, in 2015, the city launched the Sapporo Smart City Project which encouraged residents to reduce their energy consumption. Awareness campaigns, advice, and subsidies for households that introduced energy-saving or energy-recycling equipment or systems and environmental events were all focused on helping residents adopt a low-carbon lifestyle (ICLEI, 2015). In 2018, Sapporo was selected to be one of Japan's Sustainable Development Goals Future Cities, and in 2020, it achieved LEED (Leadership in Energy and Environmental Design) platinum certification with the "highest score in the world (at the time) in both the energy and water certification areas" (Biro, 2023). Sapporo has also been experimenting with the use of stored snow as a cooling system. Snow collected from road clearing is stored and then used in the summer to cool buildings. The glass pyramid inside the city's Moerenuma Park is famously cooled through snow, offsetting many tons of CO_2 emissions (Biro, 2023). Japan, along with Sweden, has been at the forefront of snow storage technology. Snow has been used since 2010 for the summer cooling of the New Chitose Airport terminals (outside Sapporo). Bibai City in central Hokkaido first experimented with the use of snow to cool data servers.

The latest initiative – Shin-Sapporo Smart City – expands on these earlier developments, trying to advance the cause of ecological sustainability and efficient energy use. Shin-Sapporo Smart City's focus is on energy savings and disaster resilience built on the Hitachi Group's natural gas cogeneration system (CGS) and Fuji Denki's AI-driven Community Energy Management System (CEMS). The Shin-Sapporo Smart City project grew out of a response to an open call for proposals by Sapporo City for innovative development projects. Daiwa House's proposal to build a low-carbon, disaster-resilient smart energy network in the Shin-Sapporo station area was selected. The goal is to make this a key part of Sapporo's development strategy and energy vision. This district of Sapporo was the site of numerous deteriorating public housing complexes that inspired the need for a development plan. On the energy side, Sapporo wanted to promote a non-nuclear energy future and a decentralised energy system. Daiwa House is leading the project in conjunction with two other construction companies (Taisei Corporation and Docon), and Hokkaido Gas, usually known as Kita Gas,[2] which supplies the energy for the project and operates the Shin-Sapporo Energy Centre which centrally manages the supply and demand of electricity and heat.[3]

Shin-Sapporo Smart City consists of seven properties. There are three speciality hospitals: the Koyukai Shin-Sapporo Hospital (185 beds), the

Shin-Sapporo Neurosurgical Hospital (135 beds), and the Shin-Sapporo Orthopaedic Hospital (88 beds). There is a medical tenancy building, a condominium building with 220 residential units, commercial facilities, and a 172-room hotel. (The hospitals opened in July 2022, the condominium in July 2023, and the commercial facilities in November 2023.) The heart of the smart city is Hitachi's natural gas CGS. CGS uses natural gas for power while the hot water and steam that are produced as a by-product supply hot water and, as there is a certain amount of demand for cooling in the summer, the exhaust heat can be used as an absorption chiller to provide air conditioning. Using the by-product of hot water and steam to produce heating or cooling makes the system particularly energy efficient. As Hitachi explains, "There is no transmission loss because power is generated as a distributed power source in the consuming region. It is also approximately twice as energy efficient as conventional thermal power generation systems that do not utilise exhaust and can use between 70 to 95% of the primary energy input into the system. This system is particularly effective at reducing CO_2 emissions because it combines the efficient use of exhaust with the low CO_2 emissions of natural gas" (Hitachi Global, 2023). During peak demand of normal operations, the CGS meets 60% of the power and 45% of the demand for hot and cold water. The rest of the required power comes from the grid, and the additional hot water comes from Hokkaido District Heating (Hitachi Global, 2023).

The Shin-Sapporo Energy Centre centrally manages the supply and demand of energy using an AI-based CEMS. The energy system aims to reduce CO_2 emissions by 35% across the smart city area (in comparison to each building having its own heat source system) using CEMS to conserve energy on both the supply and demand sides. This is done by linking energy usage information from each building and user with the Shin-Sapporo Energy Centre in real time. On the supply side, the Energy Centre endeavours to understand how and when energy is used in each facility and maximise operating efficiency by adjusting or auto-tuning equipment settings according to demand. CEMS continuously monitors the heat source equipment and automatically tunes various settings (e.g. water supply temperature, steam pressure) as demand dictates (predictive diagnostics through the use of sensors are also applied to indicate when to replace parts not based on time but based on the actual conditions of each piece of equipment). In this way, AI reduces energy use by understanding and adjusting it in real time for each facility. The hot water storage tanks in each facility (hotel, hospitals, and condominium) are connected into a single large hot water tank which allows the cogeneration waste heat to be efficiently stored and used. CEMS controls the timing of hot water storage based on when the cogeneration system will produce a surplus of exhaust heat.

On the demand side, the Energy Centre has been working with energy users (those living or working in the smart buildings) to determine how to save energy while maintaining comfort. In the common areas, the temperature can be adjusted by users within a fixed range. CEMS uses these settings to understand users' comfort levels. To enhance this knowledge, there are periodic surveys of those using the facility. For example, in November 2022, outpatients and hospital staff were interviewed about the comfort of the indoor environment including staff rooms, exam rooms, shared areas, rehabilitation area, and cafeteria. The surveys were very detailed and asked the patients and staff what they were doing in the ten minutes prior to answering the survey, what they were wearing, and whether they were sitting or standing as well as their ages and gender. In addition, if/when CEMS anticipates that energy demand will exceed supply, each facility will be requested to save electricity. Condo residents are rewarded for reducing energy demand with award points that can be used in local restaurants and stores and for gasoline from Kita Gas.

During the design phase of the Shin-Sapporo project in 2018, the importance of disaster resilience was underscored by Typhoon Jebi, which hit the Osaka region. It was the strongest cyclone to hit Japan in a quarter of a century and caused insured losses of over 12 billion USD. Typhoon Jebi was followed by the Hokkaido Eastern Iburi earthquake, which struck southern Hokkaido and left millions of residents without power. The Shin-Sapporo Smart City was designed, therefore, so that if power from the grid is interrupted during an accident, natural disaster, or maintenance on the line, electricity can be supplied from a backup line. As long as the city's natural gas continues to flow, the CGS will continue to operate and the hospitals, homes, and other facilities will continue to have heat and power. The natural gas flows through earthquake-resistant medium-pressure pipes (these were the same kinds of pipes that had survived the Hokkaido Eastern Iburi earthquake). The necessary additional heating that is supplied to the system from Hokkaido District Heating is also supplied through heat conduits made of highly earthquake-resistant steel and polyethylene pipes. Disaster resilience is a key part of the Shin-Sapporo Smart City project.

Future plans for Shin-Sapporo call for the natural gas CGS to be linked with renewable energy production outside the original Smart City block. The pipes currently used for natural gas can also carry renewable fuels like biogas and hydrogen. Future expansion could occur in multiple ways, most likely by extending the development into contiguous areas and thereby producing a larger footprint. Second, the companies and government officials behind Shin-Sapporo hope that other communities and neighbourhoods across Hokkaido and throughout Japan will be convinced to implement this new and comprehensive approach to energy management and disaster resilience.

4. Lessons from Shin-Sapporo Smart City

Although Sapporo's initiative is at the early implementation stage, it is already attracting attention due in part to the promotional efforts of the participating companies. The Smart City is, among other things, a commercial venture. The firms hope that this development, supported by the municipality and promoted as a key contribution to the country's ambitious climate change strategies, will convince other purchasers and sponsors to consider their products and services.

In this context, all climate change amelioration efforts must meet three objectives: the production of substantial and sustained reductions in energy use and carbon emissions, the demonstration of an acceptable combination of technological reliability, and the generation of enough commercial profit and user cost savings to pay for the initial and ongoing investments in technological innovation. Kita Gas and the other partner companies are attempting to meet these objectives. Cogeneration systems have been introduced to many other facilities such as the 46 Energy Centre, the Sapporo Power Station, and the Kita Gas Ishikari Power Station. At a time when corporations, governments, and institutions are making extensive and comprehensive efforts to address climate change, under the intense pressures of national and international environmental commitments, initiatives like those at Shin-Sapporo Smart City are examples of the tools available in the struggles against ecological change and future disasters.

Shin-Sapporo Smart City is one of many technological solutions to climate change requirements, an integral part of the search for financially viable and technologically sound initiatives capable of reducing a community's carbon footprint while, in this instance, improving the ability to survive major physical disasters. The world has no shortage of technological means of reducing energy use and emissions, ranging from small modular reactors (SMRs) and carbon sequestration plants to carbon air scrubbers and greatly improved building insulation. In thousands of universities, corporate laboratories, and commercial incubators around the world, research scientists, technologists, and entrepreneurs are working to identify potentially profitable and impactful technologies. The vast majority will fail to find commercial traction, at least in the short term; a few, like those being adopted in Shin-Sapporo, will move into a trial stage. Initiatives like the Shin-Sapporo Smart City endeavour to produce substantial developments in reducing energy use and carbon emissions and improving overall sustainability.

4.1 Distributed Energy System

Shin-Sapporo Smart City is an experiment about how to build a new energy model that could help Hokkaido decarbonise and will also shed light on the benefits and challenges of a distributed energy model. During the early stages of mass urbanisation, large buildings (factories, office towers, and even large apartment blocks) produced their own heat and electricity. These systems were complex, given to frequent breakdowns, expensive to maintain, and burdened by major diseconomies of scale (and the source of frequent building fires). The development of more comprehensive electrical, heating, and waste disposal systems produced major innovations in urban design, building management, energy use, and urban safety. Over time, with Japan as a world leader in imaginative and cost-effective utility services for urban areas, cities in the industrial world developed sophisticated technologies for delivering electricity and heat and for removing waste and pollutants. These cost-effective systems, producing high-quality economies of scale, were predicated on the values and priorities of the 1960s to the 1980s, including inexpensive energy, a reduction in major pollutants, and ease of delivery and maintenance.

The 21st century added a new and broader agenda, with less of an emphasis on the final cost of utilities and a greater priority on reducing emissions and protecting the urban environment from debilitating service disruptions associated with earthquakes, major fires, floods/tsunamis, and windstorms/typhoons. Japan's environmental concerns were overlaid with a focus on disaster resilience following the Kobe earthquake of 1995 and the 2011 Triple Disaster (earthquake, tsunami, and the Fukushima nuclear plant crisis). With the acceleration of natural disasters associated with climate change, the issues of utility services, ecological transition, and resilience in the face of crises emerged into sharper focus. These interwoven factors forced cities and national governments to consider new approaches to energy and utility production, all surrounded by a global commitment to reducing fossil fuel use and carbon emissions.

In the context of the historic centralisation of urban services, all predicated on capitalising on economies of scale, utility providers have been searching for new approaches that are affordable (despite expensive initial capital costs) but create lower emissions. Equally important, the search has been for decentralised approaches that, while coordinated with citywide backup systems, focus on providing energy and heat to the specific location or region. The shift has been towards decentralised utilities that provide heat and cooling, create the lowest possible carbon footprint, and have the capacity to continue operations in the face of natural or human-caused disasters. The Shin-Sapporo Smart City initiative, therefore, is one of a growing number of urban experiments, often associated with solar power or wind power installations that provide low- or no-carbon energy to specific areas while the search for citywide energy and heat production continues.

4.2 Low Carbon Emissions

On a national and international scale, the Shin-Sapporo Smart City initiative is an interesting, but not yet an enormous step towards carbon neutrality and ecological sustainability. New technologies and systems to reduce energy use and greenhouse gas emissions are underway around the world, from carbon sequestration to carbon air scrubbers, lake-based solar panels, and massive off-shore wind farms. The collective impact of these efforts, some best described as experiments but all hoping to contribute significantly to global climate change amelioration campaigns, has been limited. It is not yet clear which, if any, of a wide variety of technological innovations and scientific experiments will provide major and sustainable carbon offsets and set the world on the course of ecological sustainability.

A visit to Shin-Sapporo Smart City provides only gentle indications that the installation is connected to a broader "green" initiative. The site includes several small virtue-signalling images and colour schemes and occasional references to the energy-saving and lifestyle benefits of being connected to a unique and high-profile urban development. Despite the expense of the project and, even more, the potentially significant environmental benefits, Shin-Sapporo Smart City places as much emphasis on digital connectivity and "smart" elements as on the ecological advantages attached to the energy system. This is an indication of the developers' understanding of what is important to potential commercial clients and residents and what will help "sell" the concept to the public.

This reality speaks to the challenge associated with tackling the existential threat of climate change. To get the world back on a sustainable environmental track requires undoing a long list of major challenges: continued global population growth, high rates of energy usage, materialistic lifestyles developed over the last two centuries, an economic system that relies on unsustainable levels of energy consumption, existing ecological damage (some of which scientists have declared to be irreversible), and public expectations that the current quality of life will be improved upon in the coming years.

Many environmental commentators declare that none of these elements are sustainable, and that immediate attention is required. International conventions and consumption targets are routinely missed, which means that the environmental problems are getting worse rather than better. Thousands of demonstrations and protests have not brought about major changes in carbon production and energy usage. There is a vast literature on this, most notably the work of the Intergovernmental Panel on Climate Change. One of the most thoughtful commentators on this issue is Homer-Dixon (2010).[4] There is a massive gap between what is technically required and what is politically and economically possible. Not a single national government has, as of 2025, implemented policies that would bring about major reductions in energy use and carbon production. Japan has done more than most countries, working for decades to clean up visible and invisible environmental damage, investing heavily in electric and hydrogen vehicles, and encouraging local-level experimentation on a wide variety of energy management, recycling, and urban design fronts (Holroyd, 2018).

If anything stands out over the last two decades, it is the systematic, multi-cultural, and almost universal refusal of people to accept a major reduction in their standard of living and (in some respects) their quality of life in the pursuit of conceptual and uncertain targets for emissions reduction. When major polluters, particularly China and India, baulk at major changes in energy use, there are few incentives for countries like Japan to make more sacrifices than other nations. Practical solutions must, therefore, be economically sound and politically saleable. They have to be seen to improve environmental outcomes without undercutting citizens' quality of life and, ideally, by making companies and consumers appear to be virtuous contributors to the global climate emergency.

Japan and other leading industrial nations – heavy consumers of energy in each case – have not surrendered to the inevitability of ecological collapse. Japan has been a world leader in developing alternative sources of energy, from restarting dormant nuclear plants to exploring experimental space-based energy systems. If the source of energy can be diverted from fossil fuels to renewable energy, clearly, the country and its citizens can make major contributions to global emissions reduction and, therefore, to the battle against climate change.

The Shin-Sapporo Smart City initiative must be understood in this context, as a small but hopefully instructive experiment in the reduction of energy use and emissions production. It represents a useful element that sits at the core of the global struggle with climate change: an incrementalist, technology-based effort to produce hundreds if not thousands of initiatives that, small in themselves, nonetheless aggregate into a consequential global initiative. There does not appear to be a single invention – the ecological equivalent of the steam engine, electricity, air travel, or the Internet – that will quickly "solve" the world's current ecological dilemma. Some "moonshots", like fusion energy or SMRs, might succeed, bringing about a partial but substantial change to global energy use. A surprise technical solution – carbon scrubbers are a possible example – might bring about dramatic and substantial change. Others, like EVs, have become a focus for intense debate as to their short- and long-term contributions to mitigating climate change.

Rather than representing a single magisterial stroke to get the world out of its ecological morass, Shin-Sapporo Smart City is a small element in a comprehensive, non-coordinated, technologically sophisticated, and, ultimately, consequential effort to produce much improved environmental outcomes without a significant diminishment in current standards of living. Shin-Sapporo Smart City is not *the solution* to climate change. It is, at best, a test to see if it can be a small but significant contribution to the most complex scientific, technological, and environmental challenge in world history. It would be naïve to expect one initiative – like one based on the development of a small piece of urban real estate that uses sophisticated new technologies to reduce emissions – would address such a widespread and comprehensive threat to humanity. But it is a start, with aspirations not to address the entire climate challenge but rather to contribute (and test) a small piece of a much larger puzzle.

4.3 Energy Conservation

The first step towards carbon neutrality is energy conservation. Cogeneration is a key element in the Shin-Sapporo experiment and one of its key benefits is a reduction in energy use. The CGS uses natural gas instead of more polluting energy sources like oil or coal and, most significantly, enables a more effective use of energy by utilising what is normal waste heat. The concept is simple: maximise the energy-producing potential of any fuel consumed in supporting the operations and occupants of a building. The cogeneration system utilised in the complex uses waste heat to produce heat and hot water (and capitalising on innovation technologies, for cooling in the summer); previously, the excess heat was released into the atmosphere. By using heat that would otherwise be expelled, cogeneration saves money and energy and lessens the environmental impact of the building and its residents. What is most innovative is the use of the technologies in a large complex, producing a system that can be scaled further and replicated across the city and the country. Connecting the heating systems of a series of buildings – an area of potential energy savings and coordination that is often overlooked in urban settings – adds to the efficiency of the Shin-Sapporo Smart City.

4.4 Winter Cities

Finally, Sapporo provides another test of Japan's search for energy resilience. At the northernmost end of Japan, Hokkaido is one of the coldest parts of the country. Long and cold winters mean that the city consumes a great deal more energy each year than cities in southern climes. Energy consumption in Sapporo's winters is three to four times higher than in the summers (which, incidentally, are not as hot as Tokyo, Osaka or Fukuoka), a reality that the city shares with northern communities around the world. An innovative energy-saving urban system operating in a northern area makes a more significant contribution to national energy security and emission control than a comparable facility in a city in a warmer climate. This CGS in which waste heat is used to produce hot water and supply heat could be significant for northern cities around the world.

The Shin-Sapporo Smart City is, on a small and experimental level, a model of cooperation and coordination. The private sector proponents of the project have worked closely with the civic government and local officials to develop the initiative. Energy suppliers, technology firms, software companies, engineers, and architects designed a complex with a full understanding of the fact that it was intended to be a showcase for Japan's environmental and technological aspirations. The site itself is surprisingly low key – there are no signs in the nearby subway stations, no large billboards, and little that indicates that a high-quality and environmentally important experiment is underway.

The world is not watching Shin-Sapporo Smart City with bated breath. However, over 2,000 people (representatives from various Japanese companies, urban planners, students, academics, and the general public) visited the Shin-Sapporo New Energy Centre and the Smart City project in the first year it was open. Sapporo and Hokkaido would like to expand this model across the city and prefecture bringing down their energy use and ensuring stable access to energy in the event of a natural disaster. Perhaps other Japanese cities and foreign cities will also be interested.

For the Japanese economy, Green Growth has been a priority for several decades. Energy and emissions-savings technologies have a singular justification for saving the environment. However, climate-related developments and regulations are typically seen as antithetical to economic growth. Many observers see strong climate change initiatives as requiring a sharp decline in consumption, quality of life, and standard of living. Japan has taken a different approach, arguing that targeted investments in green technologies could produce – in addition to ecological benefits – substantial economic and employment opportunities for the country. Many recent Japanese innovations were founded on a government and business collaboration under the broad umbrella of Green Growth: EV vehicles, LED lights, rapid transit systems, Smart City energy management systems, world-leading recycling approaches, and many others. Most of these were trialed in Japan and then marketed aggressively overseas; the country's formidable international development assistance programmes have strong connections to environmental technology and the global demonstration of Japan's environmental bona fides. Japan has, for the past three decades, been at the forefront of the commercialisation of environmental technologies and the promotion of Green Growth.

5. Conclusion

It is easy to underestimate the significance of initiatives like the Shin-Sapporo Smart City. A tour of the city provides only a few small indications that this is an expensive and extensive experiment in urban climate change amelioration. It blends nicely with its surroundings in the suburbs of Sapporo and lacks the self-promotion and celebratory elements generally associated with high-profile ecological initiatives.

The challenges of climate change will not, in all likelihood, be solved by large-scale, flashy enterprises, as welcome as they might be. This existential battle will require thousands of localised, small-scale, replicable, and even scalable experiments. These projects, like Shin-Sapporo Smart City, are realistic, financially responsible, and ecologically impactful. Individually, their impacts might be small but collectively, connected to a global effort to protect quality of life while delivering real and sustainable reductions in emissions and energy use, initiatives like Shin-Sapporo Smart City may well hold the key to the long-term viability of the contemporary standard of living.

Notes

1 国家戦略特別区域法 [*kokka senryaku tokubetsu kuiki-hō*].
2 北海道ガス [*Hokkaido Gas*] or 北ガス [*Kita Gas*].
3 Material in this section is drawn from interviews and materials supplied by Ayana Okuizumi and Shingo Hagino of the Shin-Sapporo Energy Centre during a visit in May 2023.
4 See also Lawrence *et al.* (2024).

References

Biro, A. (2023) "What is a LEED Platinum City?", *gb&d Magazine*, 29 September 2023. Available at: https://gbdmagazine.com/leed-platinum-city/ (Accessed: 20 April 2024).

City of Sapporo (2013) "Sapporo City Development Strategic Vision 2013–2022". Available at: https://future-city.go.jp/en/about/ (Accessed: 20 April 2024).

Gondokusuma, M.I.C., Kitagawa, Y. and Shimoda, Y. (2019) "Smart Community Guideline: Case Study on the Development Process of Smart Communities in Japan", *IOP Conference Series: Earth and Environmental Sciences*, 294, p. 012017.

Hirayama, Y. (2021) "Japan's Smart City Initiatives Will Play Key Role in Its Digitization and Economic Revival", *Forbes Magazine*, 5 April 2021. Available at: https://forbes.com/sites/worldeconomicforum/2021/04/05/japans-smart-city-initiatives-will play-key-role-in-its-digtization-and-economic-revivial/?sh=543f25ca33d0 (Accessed: 20 April 2024).

Hitachi Global (2023) "Energy Highlights – Hitachi's Natural Gas Co-Generation System Supports the Disaster-Resilient, Low-Carbon Hokkaido Gas Energy Center". Available at: https://www.hitachi.com/products/energy/portal/case_studies/case_016.html (Accessed: 20 April 2024).

Hokkaido University (n.d.) "Sapporo Sustainability Declaration". Available at: https://global.hokudai.ac.jp (Accessed: 20 April 2024).

Holroyd, C. (2018) *Green Japan: Environmental Technologies, Innovation Policy, and the Pursuit of Green Growth*. Toronto: University of Toronto Press.

Homer-Dixon, T. (2010) *The Upside of Down: Catastrophe, Creativity, and the Renewal of Civilization*. Washington, DC: Island Press.

Hornyak, T. (2022) "Why Japan is Building Smart Cities From Scratch", *Nature*, 608, S32–S33.

ICLEI Local Governments for Sustainability, Japan Office (2015) "Low-Carbon and Resilient Cities". Available at: https://japan.iclei.org/wp-content/uploads/2021/04/catalog_en_20151222_0.pdf (Accessed: 20 April 2024).

Lawrence M., Homer-Dixon, T., Janzwood, S, Rockstöm, J., Renn, O. and Donges, J.F. (2024) "Global Polycrisis: The Causal Mechanisms of Crisis Entanglement", *Global Sustainability*, 7, pp. 1–36.

Ministry of the Environment (1989) 環境白書 [*Environmental White Paper*].

Okubo, H., Shimoda, Y., Kitagawa, Y., Gondokusuma, M.I.C., Sawamura, A. and Deto, K. (2022) "Smart Communities in Japan: Requirements and Simulation for Determining Index Values", *Journal of Urban Management*, 11, pp. 500–518.

Panasonic Group (2022) "Pioneering the Future at Tsunashima Sustainable Smart Town". Available at: https://news.panasonic.com/global/stories/1027 (Accessed: 20 April 2024).

Pham, C. (2014) *Smart Cities in Japan: An Assessment on the Potential for EU-Japan Cooperation and Business Development*. Tokyo: EU-Japan Centre for Industrial Cooperation. Available at: https://cdnw8.eu-japan.eu/sites/default/files/publications/docs/smart2020tokyo_final.pdf (Accessed: 30 April 2024).

Regan, E. (2022) "Smart Cities in Japan: Practical Innovations for Conscious Future Living", *Tokyoesque*. Available at: https://tokyoesque.com/smart-cities-in-japan/ (Accessed: 20 April 2024).

Robertson, I. (2022) "Inside Japan's Smart City of the Future", *Innovators Magazine*. Available at: https://www.innovatorsmag.com/inside-japans-smart-city-of-the-future/ (Accessed: 20 April 2024).

Sokołowski, M.M. (2021) "Models of Energy Communities in Japan (Enekomi): Regulatory Solutions From the European Union (Rescoms and Citencoms)", *European Energy and Environmental Law Review*, 30(4), 149–159.

Su, Y., Miao, Z. and Wang, C. (2022) "The Experience and Enlightenment of Asian Smart City Development: A Comparative Study of China and Japan", *Sustainability*, 14, p. 3543.

Toyota (2023) "Woven City Leaflet". Available at: https://www.woven-city.global/downloads/WovenCity_leaf_1024_en.pdf (Accessed: 20 April 2024).

United Nations Department of Economic and Social Affairs (n.d.) "News: 68% of the World Population Projected to Live in Urban Areas by 2050, Says UN". Available at: https://www.un.org/development/desa/en/news/population/2018-revision-of-world-urbanization-prospects.html (Accessed: 20 April 2024).

Wray, S. (2022) "Panasonic's New 'Smart Town' to Run on Renewable Energy", *Cities Today*. Available at: cities-today.com/Panasonics-new-smart-town-to-run-on-renewable-energy/ (Accessed: 20 April 2024).

12 Democratic Legitimacy of the Smart City

A Case Study of Yokohama

Yuichiro Tsuji

1. Introduction

Yokohama is a leading port in Japan and a core city of the Tokyo metropolitan area. The total population is approximately 3.76 million as of February 2024, the largest among Japan's cities, towns, and villages (Yokohama City, 2024). As Tokyo Metropolitan consists of special wards and municipalities, Yokohama city ranks first in population at the municipal level, followed by the city of Osaka. Yokohama has a larger population than Ibaraki Prefecture. Tokyo consists of two cities and two wards. Many private companies are located in Yokohama, and it has a high concentration of manufacturing, information technology (IT), and life science industries, including the automotive industry. It is also home to the research and development (R&D) centres of global companies such as Nissan and JGC Holdings Corporations. The city attracts numerous domestic and foreign visitors who participate in global events and explore its attractions.

In 2010, the city of Yokohama was selected as a Next Generation Energy and Social System Demonstration Area by the Ministry of Economy, Trade and Industry (METI) (Yokohama City, 2010a). Since then, the municipal government has been promoting the Yokohama Smart City Project (YSCP) demonstration project. The city aims to become a low-carbon city and plans to reduce greenhouse gas (GHG) emissions per capita by at least 60% of the 2004 level by FY2050 from April to March.

This means an annual reduction of approximately 13 million tons of CO_2, which is equivalent to more than 11,000 wind turbines and Hama Wings (Yokohama City, n.d.) in Yokohama (approximately 1,100 tons of CO_2), or a reduction of 1 kg per person per day for 3.6 million people (approximately 1.3 million tons of CO_2).

By FY2025, the city aims to reduce per capita GHG emissions by more than 30% from the 2004 levels and to increase renewable energy at the 2004 level by ten times (approximately 17 PJ). This is a reduction of about 5.3 million tons per year, which is equivalent to a reduction of more than 4,800 Hama Wing units and a reduction of 1 kg per person per day for 3.6 million people, or about four times as much.

DOI: 10.4324/9781003471448-12

Hama Wing refers to a wind farm in Yokohama, Japan. Approximately 55% of the funds for the construction of the Yokohama Wind Power Plant were raised by citizens through the purchase of "Hama Bond Windmills", a publicly offered bond issue with resident participation. The remaining 45% was subsidised by the New Energy and Industrial Technology Development Organization (NEDO). All public bonds were repaid in fiscal 2008.

The city of Yokohama has an extremely ambitious plan that will require several conditions to meet for just a local municipality to achieve this goal. Achieving this ambitious goal requires a certain population and financial size, such as Yokohama City, and its experience will influence other smaller municipalities.

1.1 Yokohama's Global Warming Action Guidelines

In 2008, the city of Yokohama established the Co-Do 30, which includes the following four basic policies (Yokohama City, 2018). Co-Do (*kodo*) means action in Japanese; it also means "code" in English. It means to change behaviour with a code (Yokohama City, 2008). First, the city has established mechanisms to reduce CO_2 emissions and improve the quality of life. Second, a focus has been placed on policy resources on effective initiatives and encouraging policy innovation by national and local governments. Third, the city has been working on actively developing market-demand pull policies. Fourth, Yokohama has been promoting initiatives through active communication and cooperation with citizens and businesses, as well as through policy coordination.

In this light, Yokohama and its citizens have been working together to create a code of conduct in the following seven areas. The first area involves changing society through individual actions to counter global warming. The second area focuses on changing society through climate-friendly businesses (product creation and services). The third area concentrates on urban development through energy-efficient buildings (equipped with energy-saving and new energy-saving devices). The fourth area involves creating attractive urban spaces where people can enjoy walking and cycling, use public transportation, and promote the de-global warming of automobiles. The city of Yokohama is promoting low-emission public buses and waste collection vehicles. The city provides financial assistance to citizens to purchase low-emission vehicles (Yokohama City, 2012). The fifth item aims to expand renewable energy tenfold. The sixth area seeks to create a green city through actions such as heat island countermeasures. Finally, the seventh agenda focuses on creating a climate-friendly city hall. As mentioned above, the linguistic coincidence in Japanese is that the word for "action" sounds identical to "code".[1] This mirrors the active policies that Yokohama is implementing to develop Yokohama-specific solutions for global warming through active collaboration between the public and the government (code), to integrate these solutions into Yokohama's practices and culture (mode).

The broad range of elements of these plans shows the complexity related to designing smart cities, namely factors such as architecture and economy, and social norms and laws that influence it. Lessig demonstrated the total relationship between norms and codes by transforming human behaviour through rewriting norms and codes. His argument would also apply to rule formation in smart cities (Lessig, 1999). To promote social norms that make citizens aware of the future vision of smart cities and implement specific actions, including those related to energy and climate, laws passed by local assemblies, which are closely linked to the lives of local residents, are preferable to laws passed by the national parliaments. In Japan, the parliament – Diet – enacts statutes binding nationally under Article 41 of the Constitution of Japan and the local governments pass ordinances under Article 94 of the Constitution.

1.2 Forming Standards Through Planning

The city of Yokohama's Global Warming Action Plan is a strategy for local governments to reduce GHG emissions under Article 21(3) of the Act on Promotion of Global Warming Countermeasures (1998). This law authorises local governments to establish plans for measures to reduce GHG emissions in their affairs and projects, either individually or jointly, in line with the global warming action plan.

One of Yokohama's measures to combat the negative consequences of climate change is a plan to realise a state-of-the-art smart city. Planning is a classic instrument of public management existing in different policy areas. For example, Article 40 of the Basic Act on Disaster Management (1961) provides that the central government shall prepare a basic plan for disaster reduction, and each local government shall establish its own disaster reduction plan.

These instruments, however, differ globally, also in terms of the relationship between the central and local governments. Generally, in Japan, the relationship between central and local power is related to the fact that in many areas local governments are responsible for implementing policies formulated by the central government under the Local Autonomy Act. One of the most prominent textbooks of the Japanese Constitution explains that the central government with specific ministries is concerned with enacting laws and creating budgets and tax systems, while local governments are actually required to operate within the legal framework (Colin *et al.*, 2023). Therefore, local governments are obliged to make specific plans to implement the law.

For the affairs to be handled, under ex-Article 2(4) of the Local Autonomy Act, a local government used to try to clarify the goals and means of administrative management in three layers: a basic concept, a master plan, and an implementation plan. Ex-Article 2(4) provided that, in handling its affairs, a municipality shall establish a basic plan for comprehensive and systematic administrative management in its area through a resolution of its council and shall ensure that its affairs are conducted in accordance with such plan.

All could be classified as administrative plans. Their origin could be traced back to the Local Autonomy Act, a legislation that mandated local governments to formulate such plans. However, the ex-Article 2(4) was repealed in 2011. Then, the Local Autonomy Act was partially amended and the obligation to prepare a basic plan was abolished. Since local governments are not legally obligated to enact basic concepts, each local government was left to come up with its own system of basic concepts and plans (Okitsu, Y., 2023).

Nevertheless, many local governments in Japan still use comprehensive plans as the top-level plans for administrative operations. With the repeal of ex-Article 2(4) of the Local Autonomy Law, cities are no longer legally obligated to formulate a basic concept and are left to their own judgement as to whether or not to formulate and go through the resolution of the local assembly. The Minister of Internal Affairs and Communications (2011) issued a notice stating that individual municipalities may continue to formulate the basic concept at their own discretion and through a resolution of the local assembly. Based on this notice, which is a guideline for interpretation of the Local Autonomy Act, an increasing number of local governments are formulating their basic concepts based on ordinances.

Such a plan outlines the future goals and policies of the local government shared by all residents, and comprehensive plans are adopted by city and town assemblies, taking into account the will of the residents and, in many cases, after deliberation by the assemblies. The comprehensive plan is the basic guideline for all residents, businesses, and governments to act on.

The basic concept was considered the highest level of planning and was formerly required by the Local Autonomy Act for municipalities to formulate administrative plans. In Japan, the Local Autonomy Act was revised in 2011 amidst reforms to address the strong authority of the central government that should be transferred to local governments, and the provision requiring the formulation of the basic concept was abolished. However, some homepages of local governments still consider the basic concept at the top level.

Therefore, the Basic Policies for Economic and Fiscal Management and Reform, which provides a general direction for economic and fiscal policy in 2022, expressed the idea of minimising the number of administrative plans that the government requires local governments to formulate. By reducing the central government's involvement in local affairs, the Basic Policy aims to reduce the burden on local governments and increase their discretion (Cabinet Office, 2021).

This framework shows the legal regime in which the Japanese plan for smart cities is placed. In general, it is based on a system in which the central government (METI) gives high marks to the proactive efforts of local governments. METI provides financial support for the high-marked projects. One of them is the YSCP. The legal obligation under Articles 19 to 22–4 of the Act on Promotion of Global Warming Countermeasures (1998), which requires local governments to prepare plans, has the effect of promoting competition among local governments by encouraging each to implement plans.

A key factor that affects the feasibility of a smart city plan is the size of the municipality's population and finances, as some municipalities may not have sufficient human resources and expertise to prepare and implement such plans. The city of Yokohama, being an ordinance-designated city under Article 252–19 of the Local Autonomy Act – which provided that a city is as categorised as ordinance-designated city if the population of the city is 500,000 or more as designated by a Cabinet Order – has the advantage of receiving various operations and new financial resources from the prefectural government, which enables it to offer advanced, specialised, and tailored administrative services to its citizens. Therefore, Yokohama, as an ordinance-designated city with a large population, has the potential to attract many businesses to participate in its smart city plan. Moreover, it can provide a valuable example for other municipalities to follow (Farzaneh *et al.*, 2014).

2. The City of Yokohama's Smart City Project

The city of Yokohama has successfully implemented smart city initiatives related to sustainable urban development in cooperation with citizens and businesses. To create a plan, its democratic legitimacy requires the proactive participation of residents. Only when the city provides residents with sufficient information can they become interested in and proactively participate in smart city planning.

To handle matters under the Local Autonomy Act, local governments often establish a basic concept for comprehensive and systematic administrative management in the area through a resolution of the local assembly. These plans are collectively referred to as the "comprehensive plan". A comprehensive plan is also called a "master plan" because of its wide range and strategic character.

Local governments are not obligated to formulate a basic concept under the Local Autonomy Act, and many prefectures prepare comprehensive plans under the name of master plans. The master plan for the YSCP (Yokohama City, 2010a) is positioned within this framework.

2.1 Participation of Residents in Planning: Yokohama Smart Business Association

Under Article 94 of the Constitution of Japan and Article 14 of the Local Autonomy Act, democratic legitimacy can be ensured by providing opportunities for residents to participate in the preparation of comprehensive plans by municipalities. It is precisely because the voices of residents are reflected in the plan that legitimacy is created in the plan. Local governments identify the needs of residents in defining the basic concept by soliciting questionnaires, opinions, and suggestions from the public and may make drafts of the basic concept publicly available to residents to obtain their opinions. Residents may establish a study organisation composed of residents, academics, and experts.

There are several benefits of residents' involvement in administrative planning. First, local governments can capture the nuanced needs and views of residents and incorporate them into the comprehensive plan. Questionnaires alone are limited by the preset questions and choices, leaving out free-answer columns in Japan. They also fail to convey to the respondents how their opinions are considered in the plan. Second, residents' creative ideas can enrich administrative planning. In Japan, over the years, the suggestions from the individuals and organisations have improved and been of high quality and policy-oriented. An important step to facilitate this process was when after the Great Hanshin Earthquake in 1995 the Act on Promotion of Specified Non-profit Activities (1998) was enacted.

Since there was no legal system to register a non-governmental organisation (NGO) in Japanese society until the Great Hanshin Earthquake, simply starting as an activity makes it a voluntary organisation. In order to start up as a corporate organisation, it is necessary to be certified as a non-profit organisation (NPO). The role of NGOs has had a significant impact on the enactment of local government ordinances.

Third, involving residents in planning can enhance their awareness and engagement in creating climate-friendly solutions for smart cities. Local government plans in Japan are often too vague to be fully appreciated by residents, and sometimes even by governmental officials. The basic plan drafted by the local government is a fundamental notion, but it is general, and thus it demands a mechanism for residents to share a high level of understanding and awareness of the issues. Basic plans and implementation plans are more concrete than the basic concept, but they are overloaded with information. By engaging residents in the planning process, they will sense that their views are reflected in the plan. The governmental officials will also foster a spirit of cooperation and collaboration with the residents through their organised support.

In this context, the city has established the Yokohama Smart Business Association (YSBA) (Yokohama City, 2010a), which seeks to achieve its goals using low-carbon related technologies in three areas: energy, buildings, and transportation. YSBA is a basis for ensuring democratic legitimacy and is a partnership between the public and private sectors. Based on the YSCP demonstration (practical) project, this is a plan to test technologies, systems, and institutions that are at a stage where they can be applied on a practical basis, and to confirm their effectiveness and economic efficiency, up to FY 2014.[2] YSBA was established in 2015 to develop the project from "demonstration to implementation" by utilising the technology and know-how cultivated in the YSCP, aiming to create an energy-recycling city with excellent disaster prevention, environmental, and economic performance. This corresponds to the three-step approach implemented in Yokohama: first, deploying the cutting-edge technologies in an integrated platform to showcase their performance; second, testing the economic viability of the service models and new businesses that can facilitate the adoption of the validated technologies; and third, evaluating the social impact of the diffusion of the established services, along with the institutional design that supports them.

In this light, the city of Yokohama has been working on a project to review practicability of YSCP; for this reason, there have been collaborations with 34 Japan's leading energy-related businesses, ranging from electrical manufacturers to and construction companies. These efforts have been aimed at installing systems to optimise energy supply and demand in homes, business buildings, and other urban areas. Moreover, Yokohama, in cooperation with YSBA, has set the goal of establishing a social system that utilises low-carbon technologies in three areas: energy, buildings, and transportation and traffic, to reduce CO_2 emissions on a large scale.

2.2 YSCP Master Plan

In 2010, the city of Yokohama formulated the YSCP Master Plan with the goal of realising a low-carbon city to address the climate challenges it faces, such as countering global warming. It was also selected as one of the demonstration areas in the Next Generation Energy and Social System Demonstration Project (METI). In terms of chronology, it was YSCP 1.0. Sixteen projects were selected as components of the smart city, including the home (HEMS, Home Energy Management System), building (BEMS, Building Energy Management System), (EV, Electric Vehicle) charging station, storage battery (SCADA, Supervisory Control and Data Acquisition), and CEMS (Community Energy Management System) that oversees the entire system (Yokohama City, 2010b).

In 2015, YSCP 2.0 was established. It was a new public–private partnership organisation, the already-mentioned YSBA, to manage activities ranging from demonstration to implementation of EMS (Energy Management System) in three regions of Minato Mirai 21, Pacifico Yokohama, and Tsurumi by utilising technologies, know-how, and other results developed in the YSCP demonstration project. For instance, the YSBA conducted a feasibility study – subsidised by the METI – that used energy analysis and evaluation to design a low-carbon city plan for Yokohama. The report served as a master plan to foster local energy production and consumption. The YSBA has been pursuing the vision of an energy-recycling city that boasts outstanding environmental, disaster prevention, and economic outcomes. The city's goal is to become a model of energy recycling that is environmentally friendly, disaster-resilient, and economically efficient.

In 2019, YSBA 3.0 started, and in February 2023, the city of Yokohama released the Yokohama Smart City Master Plan (YSCP 3.4). The YSCP 3.4 Master Plan (Yokohama City, 2023) is part of the Sustainable Development Goals (SDGs), namely the Future City Plan and the Yokohama City Global Warming Prevention Action Plan, and will utilise the various urban facilities, electric vehicles, independent and distributed power sources, etc. that exist in the city, as well as efforts to lower CO_2 emissions. The goal of the action plan is to contribute to the realisation of the SDGs and Zero Carbon Yokohama by creating a state-of-the-art smart city for decarbonisation. YSCP 3.0 Master Plan covers 2019–2023, and has been reviewed, revised, and renewed. The total budget for FY 2021 was 18,915,000 JPY (Yokohama City, 2022).

The YSCP 3.4 Master Plan included the following seven elements. Table 12.1 highlights them in detail (Yokohama City, 2023).

Table 12.1 Core elements of the YSCP 3.4 Master Plan.

Project name	*Details*
EV Charging Infrastructure Expansion Project	Yokohama City and e-Mobility Power Co. Ltd. have concluded a partnership agreement to expand the development of recharging infrastructure to promote the spread of EVs. In addition, EVs can also be used to supply electricity in times of disaster, etc.
High-Efficiency Heat Source Manufacturing Equipment at Yokohama Station West Exit District Heating and Cooling Facility	Yokohama Station West Exit District Heating and Cooling Facility Ltd., which operates a heat supply business in the area around the west exit of Yokohama Station, will undertake a major upgrade of its heat source equipment. This is being added as a new project because it will make a significant contribution to reducing CO_2 emissions in the area. This project was added in YSCP 3.1.
Renewable Energy in City Facilities	In Yokohama, private companies are installing solar power generation systems and battery storage systems at elementary and junior high schools in the city. Yokohama purchases the renewable energy electricity generated by the facilities to cover the installation and management costs of the facilities. This project is included as a new project because it contributes significantly to the reduction of CO_2 emissions.
GTL Fuel Utilisation Project (This project was added in YSCP 3.1.)	Reduce CO_2 emissions from construction work in Yokohama City by Itochu Enex Corporation utilising GTL fuel. This company is a new member of YSBA.
Switching to Renewable Energy Electricity (This project was added in YSCP 3.2)	This project will contribute to the reduction of CO_2 emissions of businesses by switching to 100% renewable energy through a CO_2-free rate menu that combines electricity generated by renewable energy and non-fossil certificates designated as renewable energy in our electricity contracts, targeting environmentally advanced businesses and households in the city.
Methanisation Project	This project of Tokyo Gas Co. Ltd. and the city of Yokohama aims to build a decarbonisation model of local production for local consumption by promoting the demonstration of carbon-neutral methane production, etc. in conjunction with renewable energy.
Renewal ZEB Model Demonstration Project	Taisei Corporation will renovate and convert an existing building, the Yokohama Branch Building, into a ZEB building by maximising energy conservation and renewable energy, and by using cutting-edge, highly versatile technology.

As of February 2023, when the YSCP 3.4 was published, 19 projects had been conducted. Each business can test its own services with the citizens of Yokohama, as long as they align with the goals of the YSBA. These smart city projects are designed to be platforms, with the feature that the more users they have, the lower the service cost and the user fees. Since these are trial projects, they will foster a strong bond between the citizens of Yokohama and the companies that joined the experiment, if they are successful.

3. Legitimacy of the Council and Citizen Participation

The YSBA plays an integral role in the design of the administrative plan for the Smart City of Yokohama. In general, in Japan, councils can be legally classified into three categories. First, they are councils established as an auxiliary body, similar to an advisory board. Second, there are councils formed among local governments or between governments such as local governments and the national government. Examples of this type of council are the recovery council under Article 11(1) of the Act on Recovery from Major Disasters (2013), or the urban renewal and emergency improvement council under Article 19(1) of the Act on Special Measures Concerning Urban Reconstruction (2002). These councils deliberate on the planning and execution of major disaster recovery and urban development. Third, there are councils organised between the national or local government and the private sector. The YSBA belongs to this category.

The YSBA (Yokohama City, 2021) consists of 8 executive committee members and 17 general members. The city of Yokohama is recruiting members for the council. To become a member of the YSBA, a company or organisation must agree with the objectives of the YSBA and be willing to participate in the promotion of its projects. Organisations willing to become members should submit an application form to the YSBA's Secretariat. The application must be accompanied by a letter of recommendation for membership from members of the Executive Committee. The Executive Committee decides whether the prospective organisation is eligible for membership. The YSBA executive membership includes Azbil Corporation, Taisei Corporation, Tokyo Gas Company, Tokyo Electric Power Energy Partners, Toshiba Energy Systems Corporation, Minatomirai 21 Heat Supply Company, Meidensha Corporation, and the city of Yokohama. YSBA's general membership includes many large corporations, such as IHI Corporation, e-Mobility Power Corporation, Itochu Enex Corporation, Orix Corporation, Shimizu Corporation, Takasago Thermal Engineering Corporation, Tokyo Gas Engineering Solutions Corporation, and Tokyo Urban Services Corporation.

3.1 Other Councils and YSBA

The main objective of the YSBA is to utilise and further develop the knowledge gained from the YSCP demonstration experiment and to flexibly adapt to changing energy-related services in order to decarbonise the urban region. The YSBA will exchange information and discuss the following issues related to the operation of the public–private partnership platform: first, to promote local production for local consumption of energy and to improve the efficiency of energy use and disaster prevention; second, to stimulate the economy by creating new services based on the deregulation of electricity and gas retailing; third, to further increase public awareness of measures to combat global warming; and fourth, to work to achieve decarbonisation in other parts of Yokohama.

In addition to the YSBA, the following councils exist where local government and private citizens (private companies) work together to make decisions relevant to climate and environment. This is, for instance, a council organised between a local government and a private operator (the third type) as provided in the Act on Promoting Generation of Electricity from Renewable Energy Sources Harmonized with Sound Development of Agriculture, Forestry and Fisheries (2014). The purpose of this Act is to revitalise agricultural, mountainous, and fishing villages by taking measures to promote the generation of electricity from renewable energy sources in harmony with the sound development of agriculture, forestry, and fisheries, and to establish a system that contributes to the diversification of energy sources.

The Act on Revitalization and Rehabilitation of Local Public Transportation Systems (2007) established this type of council as well. According to Article 6(1) of the Act, the purpose of this Act is to obligate local governments to play a proactive role in maintaining public transportation in the region and to create a system that ensures transportation for daily life.

Municipal Urban Planning Councils under the Act on Special Measures Concerning Urban Reconstruction (2002) is another example. The purpose of this law is to take comprehensive measures to control new settlements in disaster-prone areas, promote relocation, and promote disaster-resistant urban development to cope with the increasing frequency and severity of natural disasters.

This type of council attempts to strengthen democratic legitimacy by allowing the voices of local residents to be reflected in ordinances through the council. This is relatively common in laws enacted after 2000. If it is an annex to the executive body of a local government, it is sufficient to have a statutory provision and no further ordinance is required under Article 202–3 of the Local Autonomy Act – this Act provides that an auxiliary organ of an executive organ of an ordinary local public entity shall be an organ that conducts conciliation, examination, deliberation, or investigation, etc. on the matters it is in charge of, pursuant to the provisions of laws, Cabinet orders based thereon, or ordinances. For councils required by law, additional ordinance may be established. Some statutes require that the council be provided for by law and further create provisions in the ordinances. Under the Special Measures Law for Nuclear Evacuees (2011),[3] a designated municipality may, by ordinance, establish an address relocation council under this Act. Under the Revised Bicycle Law (1981),[4] a municipality may, in accordance with the provisions of its ordinance, establish a Bicycle and Other Motor Vehicle Parking Council to study and advie on important matters relating to parking policies for bicycles and other motor vehicles. The Basic Act on Traffic Safety Measures (1970) law permits prefectures to establish a prefectural traffic safety liaison council, as specified by ordinance, when deemed necessary for liaison and consultation with relevant local administrative agencies regarding the safety of marine or air traffic in the prefecture's area.

3.2 Democratic Legitimacy of Councils

The reason for stipulating the council in the ordinance seems to be to secure democratic legitimacy, even though the Japanese law does not mandate the creation of an ordinance. In general, these councils would require citizen involvement. By law, smart city councils like the YSBA do not need to create ordinances, but can set up terms of reference[5] to ensure appropriate operation and legal certainty, in response to the demand of private sector actors.

For example, Article 8 of the Act on Special Measures Concerning Appropriateness and Revitalization of General Passenger Vehicle Transportation Business in Specified and Quasi-Specified Regions (2009)[6] does not require that the Council establish terms of reference for its organisation, but the basic policy under the Law does (MLITT, 2009). According to Article 8–2 of this Law, regional councils establish regional plans and work to ensure proper management of taxi services. The regional plan shall address the revitalisation of taxi services, the revitalisation and efficiency of business management, the prevention of deterioration of working conditions for taxi drivers, the improvement and enhancement of working conditions, and the improvement of traffic, environmental, and urban problems. The taxi operators shall prepare a specific business plan to realise the optimisation of business in accordance with this regional plan and obtain the approval of the Minister of Land, Infrastructure, Transport, and Tourism.

Through these councils, various parties can participate in the preparation of the plan to reflect their different interests and opinions. In preparing the administrative plan, it is necessary to maintain a certain distance from business operators.

The legitimacy of collaborative decision-making between government and private citizens is not an issue unique to Japan. Too close a distance between the city and the business community might inhibit competition. Stewart (1975) has already pointed out that in administrative decision-making, the opinions of the members are gradually emphasised and fail to reflect the opinions of society as a whole.

While businesses exercise their constitutional right under Article 22 of the Constitution of Japan to freely engage in economic activities, as members of the council, their economic activities may be restricted by decisions of the council. If the council were to balance the interests of companies, the decisions of the council could work to the detriment of companies that do not participate in the council.

In this regard, Freeman (2000) suggests that in the council, administrative plans are decided through negotiations between local governments and companies. There is some debate as to whether this process is viewed favourably or negatively. Some councils can potentially make democratic decisions obsolete. In terms of administrative plans, in a public–private council, the coordination of interests is not done by the government agency, which is the planning body, but by the council, of which the parties concerned are also members.

Easterbrook (1984) suggests if a private organisation seeks to achieve its own interests through negotiation without regard to the interests of others, i.e. the public interest, he questions how the judiciary can uphold such an administrative decision.

One should note that the plan formulation might be influenced by large corporations, and not represent the views of small individuals. This appears to be a problem in Yokohama, where the YSBA reflects the voices of large corporations more than those of Yokohama's citizens. This phenomenon is not unique to Japan. Dave O. (2022) argues as follows:

> [m]easuring the extent to which these problems occur is difficult, but both regulators and attorneys representing regulated parties agreed that smaller, less resourced entities face disadvantages in negotiation-based systems. As one regulator explained, in response to a question about whether it is easier for larger companies to navigate regulatory negotiations: "[i]t absolutely is. They have more resources and more capacity to deal with developing a plan while maintaining their operations".

Therefore, the YSBA should operate in a way that does not favour only those companies, but also considers the needs of the public and other companies that do not join the council. With their economic power and advanced technological capabilities, large corporations can accelerate the development of smart cities in Yokohama. However, this requires more transparency that would complement the accountability of the local government of Yokohama.

4. Conclusion

The idea of the smart city in Yokohama is based on a goal to tackle global warming under the Act on the Promotion of Global Warming Countermeasures (1998). The YSCP has been largely successful. This was possible due to its large population, financial strength, and authorities willing to collaborate with the prefectural government. Another factor for success is the private sector cooperation with the municipal government. As a result, the YSCP 3.0 Master Plan enables the realisation of a state-of-the-art smart city by utilising the elements of the YSCP, such as promotion of electric transportation, independent and distributed power sources, and so on. In effect, the city of Yokohama has expanded its EV charging infrastructure, installed high-efficiency heat generation equipment at the district heating and cooling plants, created renewable energy in city-owned and -managed facilities, and has been focused on promoting the transition to renewable energy.

An institutional facilitator of public–private cooperation is the YSBA, a council composed of the representatives of the city of Yokohama and large private businesses. Through the active participation of the private sector, it is the key institution to promote smart and climate-friendly urban solutions. Its problem is, however, the lack of democratic legitimacy of the YSBA alone in terms of reflecting the voices of Yokohama's citizens in the design of the smart city. To maintain the legitimacy of the smart city planning, priority must be given to ensuring that the benefits of the administrative plan accrue to the communities and its citizens most affected by smart city development. There are many benefits of active participation of residents in the planning process. Yokohama should be more open to residents' ideas that could be incorporated into the planning process. Increased awareness and participation in the smart city planning process is a win-win option.

Notes

1 An "action" in Japanese is 行動 [*kodo*], and "code" in Japanese is コード [*kodo*].
2 The name of the version of YSCP depends on the time of implementation. It is YSCP 1.0 from 2010 to 2014. (Yokohama, 2020b).
3 Full name: Act on Special Provisions for Administrative Procedures Pertaining to Evacuees and Measures Pertaining to Persons Relocating Their Addresses in Response to Disasters Caused by the Accident at Nuclear Power Plants in the Great East Japan Earthquake.
4 Full name: Act on Promotion of Safe Use of Bicycles and Comprehensive Advancement of Measures for Bicycle Parking.
5 In Japanese: 規約 [*kiyaku*].
6 Also known as the Taxi Appropriateness and Revitalization Law.

References

Act on Promoting Generation of Electricity From Renewable Energy Sources Harmonized With Sound Development of Agriculture, Forestry and Fisheries (2014) No. 81. In Japanese: 農林漁業の健全な発展と調和のとれた再生可能エネルギー電気の発電の促進に関する法 [*nōrin gyogyō no kenzen-na hatten to chōwa no toreta saisei kanō enerugi denki no hatsuden no sokushin ni kansuru hō*]. Available at: https://laws.e-gov.go.jp/law/425AC0000000081 (Accessed: 3 January 2025).

Act on Promotion of Safe Use of Bicycles and Comprehensive Advancement of Measures for Bicycle Parking (1981) No. 87. In Japanese: 自転車の安全な利用の促進及び駐車場対策の総合的推進に関する法 [*jitensha no anzen-na riyō no sokushin oyobi chūshajō taisaku no sōgōteki suishin ni kansuru hō*]. Available at: https://laws.e-gov.go.jp/law/355AC1000000087 (Accessed: 3 January 2025).

Act on Promotion of Specified Non-Profit Activities (1998) No. 7. In Japanese: 特定非営利活動促進法 [*tokutei hi-eiri katsudō sokushin hō*]. Available at: https://laws.e-gov.go.jp/law/410AC1000000007 (Accessed: 3 January 2025).

Act on Recovery from Major Disasters (2013) No. 55. In Japanese: 大規模災害からの復興に関する法 [*daikibo saigai kara no fukkō ni kansuru hō*]. Available at: https://laws.e-gov.go.jp/law/425AC0000000055/ (Accessed: 3 January 2025).

Act on Revitalization and Rehabilitation of Local Public Transportation Systems (2007) No. 59. In Japanese: 地域公共交通の活性化及び再生に関する法律 [*chiiki kōkyō kōtsū no kasseika oyobi saisei ni kansuru hō*]. Available at: https://laws.e-gov.go.jp/law/419AC0000000059 (Accessed: 3 January 2025).

Act on Special Measures Concerning Appropriateness and Revitalization of General Passenger Vehicle Transportation Business in Specified Areas (2009) No. 64. In Japanese: 特定地域及び準特定地域における一般乗用旅客自動車運送事業の適正化及び活性化に関する特別法 [*tokutei chiiki oyobi jun tokutei chiiki ni okeru ippan jōyō ryokyaku jidōsha unsō jigyō no tekiseika oyobi kasseika ni kansuru tokubetsu hō*]. Available at: https://laws.e-gov.go.jp/law/421AC0000000064 (Accessed: 3 January 2025).

Act on Special Measures Concerning Urban Reconstruction (2002) No. 22. In Japanese: 都市再生特別措置法 [*toshi hukkō tokubetsu sochi hō*]. Available at: https://laws.e-gov.go.jp/law/414AC0000000022/ (Accessed: 3 January 2025).

Act on Special Provisions for Administrative Procedures Pertaining to Evacuees and Measures Pertaining to Persons Relocating Their Addresses in Response to Disasters Caused by the Accident at Nuclear Power Plants in the Great East Japan Earthquake (2011) No. 98. In Japanese: 東日本大震災における原子力発電所の事故による災に二対処するための避難民にかかる事務処理の特例及び住所移転者にかかる措置に関する法 [*higashi nihon daishinsai ni okeru gensiryoku hatsudensho no jiko niyoru saigai ni taisho surutameno hinan jumin ni kakaru jimushori no tokurei oyobi jūsho itensha ni kakaru sochi ni kansuru hō*]. Available at: https://laws.e-gov.go.jp/law/423AC0000000098/ (Accessed: 3 January 2025).

Act on the Promotion of Global Warming Countermeasures (1998) No. 117. In Japanese: 地球温暖化対策の推進に関する法 [*Chikyū ondanka taisaku no suishin ni kansuru hō*]. Available at: https://laws.e-gov.go.jp/law/410AC0000000117/ (Accessed: 3 January 2025).

Basic Act on Disaster Management (1961) No. 223. In Japanese: 災害対策基本法 [*saigai taisaku kihon-hō*]. Available at: https://laws.e-gov.go.jp/law/336AC0000000223/ (Accessed: 3 January 2025).

Basic Act on Traffic Safety Measures (1970) No. 110. In Japanese: 交通安全対策基本法 [*kōtsu anzen taisaku kihon hō*]. Available at: https://laws.e-gov.go.jp/law/345AC0000000110 (Accessed: 3 January 2025).

Cabinet Office of Japan (2021) "Basic Policy on Economic and Fiscal Management and Reform". Available at: https://www5.cao.go.jp/keizai1/basicpolicies-e.html (Accessed: 19 April 2024).

Colin, J. (2023) *The Annotated Constitution of Japan: A Handbook*. Amsterdam: Amsterdam University Press.

Dave, O. (2022) "The Negotiable Implementation of Environmental Law", *Stanford Law Review*, 75(1), pp. 137–203.

Easterbrook, F.H. (1984) "The Supreme Court, 1983 Term – Foreword: The Court and the Economic System", *Harvard Law Review*, 98(1), pp. 4–60.

Farzaneh, H., Suwa, A., Dolia, C.N.H. and de Oliveira, J.A.P. (2014) "Developing a Tool to Analyze Climate Co-Benefits of the Urban Energy System", *Procedia Environmental Sciences*, 20, pp. 97–105.

Freeman, J. (2000) "The Private Role in Public Governance", *New York University Law Review*, 75(3), pp. 543, 653–661.

Lessig, L. (1999) "The Law of the Horse: What Cyberlaw Might Teach", *Harvard Law Review*, 113(2), pp. 501–549.

Local Autonomy Act (1947) No. 67. In Japanese: 地方自治法 [*chihō jichi hō*]. Available at: https://laws.e-gov.go.jp/law/322AC0000000067#Mp-At_4 (Accessed: 19 April 2024).

Minister of Internal Affairs and Communications of Japan (2011) Notice No. 51 on Partial Amendment of the Local Autonomy Act.
Ministry of Land, Infrastructure of Japan (2009) Notice No. 1036 – Transport and Tourism, Basic Policy on the Appropriateness and Revitalization of General Passenger Vehicle Transportation Business in Specified and Quasi-Specified Regions. In Japanese: 特定地域及び準特定地域における一般乗用旅客自動車運送事業の適正化及び活性化に関する基本方針 [*tokutei chiiki oyobi jun tokutei chiiki ni okeru ippan jōyō ryokaku jidōsha unsō jigyō no tekisei-ka oyobi kassei-ka ni kansuru kihon hōshin*]. Available at: https://wwwtb.mlit.go.jp/tohoku/content/000180889.pdf (Accessed: 3 January 2025).
Okitsu, Y. (2023) 行政法 [*Administrative Law*]. Tokyo: Shinseisha.
Stewart, R.B. (1975) "The Reformation of American Administrative Law", *Harvard Law Review*, 88(8), pp. 1667–1813.
Yokohama City (2008) "CO-DO 30". Available at: https://www.city.yokohama.lg.jp/shikai/kiroku/katsudo/h21-h22/katsudogaiyo-h21-t-4.files/0009_20180808.pdf (Accessed: 19 April 2024).
Yokohama City (2010a) "Yokohama Smart City Project". Available at: https://www.city.yokohama.lg.jp/kurashi/machizukuri-kankyo/ondanka/etc/yscp/yscp01.files/YSCP_MP.pdf (Accessed: 19 April 2024).
Yokohama City (2010b) "Announcement of City of Yokohama Project Master Plan". Available at: https://www.city.yokohama.lg.jp/kurashi/koseki-zei-hoken/zeikin/shizeikanren/chosakai/jisseki/kaigisiryou0.files/0235_20240730.pdf (Accessed: 19 April 2024).
Yokohama City (2012) "Yokohama City Automobile Pollution Prevention Plan FY2012–2017". Available at: https://www.city.yokohama.lg.jp/kurashi/machizukuri-kankyo/kankyohozen/hozentorikumi/kotsukankyo/jikobo.files/0001_20180820.pdf (Accessed: 19 April 2024).
Yokohama City (2018) "Yokohama Action Policy on Climate Change". Available at: https://www.city.yokohama.lg.jp/kurashi/koseki-zei-hoken/zeikin/shizeikanren/chosakai/jisseki/zeiseikenkyukai/10kenkyukai.files/0007_20180717.pdf (Accessed: 19 April 2024).
Yokohama City (2020a) "Yokohama City Action Plan for Global Warming Countermeasures". Available at: https://www.city.yokohama.lg.jp/lang/overseas/climatechange/contents/zcy/actionplan.html (Accessed: 19 April 2024).
Yokohama City (2020b) "Yokohama Smart City Project (YSCP)". Available at: https://www.city.yokohama.lg.jp/lang/overseas/climatechange/contents/energypolicy/yscp.html (Accessed: 19 April 2024).
Yokohama City (2021) "Recruitment of Members for the Yokohama Smart Business Council". Available at: https://www.city.yokohama.lg.jp/business/bunyabetsu/kankyo-koen-gesui/ondanka/YSCP06.html (Accessed: 19 April 2024).
Yokohama City (2022) "2022 Business Valuation Report". Available at: https://www.city.yokohama.lg.jp/city-info/gyosei-kansa/innovation/jigyohyoka/r04/reiwa03jigyo.files (Accessed: 19 April 2024).
Yokohama City (2023) "YSCP 3.4 Master Plan". Available at: https://www.city.yokohama.lg.jp/kurashi/machizukuri-kankyo/ondanka/etc/yscp/yscp05.files/0042_20230222.pdf (Accessed: 3 January 2025).
Yokohama City (2024) "Statistics Portal". Available at: https://www.city.yokohama.lg.jp/city-info/yokohamashi/tokei-chosa/portal/ (Accessed: 3 January 2025).

13 Sustainable Smart City Tokyo

Between Problems of the Past and Chances of the Future

Maciej M. Sokołowski and Fumio Shimpo

1. Introduction

Tokyo is a city that attracts. The modern capital of Japan has long had the status of a global brand – a recognisable symbol of modernity, good organisation, development, culture, sport, or fashion. It is also one of the world's largest cities with a population of about 14.06 million, 11% of Japan's total, that has a gross metropolitan product reaching about 110 trillion JPY,[1] which is 20.4% of Japan's gross domestic product, and manages own budget accounting for 16 trillion JPY (Tokyo Metropolitan Government, 2023, p. 5).[2]

> To the visitor, Tokyo is a city on the move. High-rise buildings are changing its skyline. Offices are modern. Traffic flows along gracefully winding freeways. Its people are better dressed than ever before. Shop windows are full of attractive merchandise. Neon signs beckon invitingly over entertainment districts. Hotels are usually full. But behind this façade, Tokyo is in trouble.
>
> True, the people of Tokyo, like the people of the rest of Japan, enjoy the highest standard of living in their history. But the rapid pace of economic expansion has created numerous distortions and imbalances. In the race to build more factories and produce more goods, the welfare of the needy, who could not afford the new material luxuries, was given little attention. In addition to producing more goods, the concentration of factories has produced air pollution and water contamination, both serious dangers to health.
>
> (Tokyo Metropolitan Government, 1969a, pp. 10–11)

Because of its status, Tokyo has been, is, and will continue to be a benchmark, not only for other Japanese cities but also for cities in Asia and the world. However, is Tokyo a smart city? If we refer to the Japanese Smart City Guidebook (Cabinet Office, 2021), we will find that although smart cities have different forms – they generally *provide*, *enhance*, and *solve*. First, they provide "services to support each one of residents using new technologies"; second, they enhance "management in various fields (e.g. planning, development,

DOI: 10.4324/9781003471448-13

management/operation)"; and third, they solve "challenges faced by cities and regions", and continue "to create new value" (Cabinet Office, 2021, p. 9). Moreover, these cities must be "sustainable" and must realise the idea of Society 5.0. "ahead of the others" (see Cabinet Office, 2021, p. 9).[3]

Given this background, in this chapter, we analyse Tokyo in the context of its smart city policies, specifically focusing on its sustainability. However, acknowledging that Tokyo, like Rome, "was not built in a day", we begin our analysis by examining the environmental challenges the city has faced historically. This discussion covers issues such as pollution and contamination in Tokyo during the 1960s and 1970s, with a particular emphasis on air quality. Following this, by reviewing Tokyo's strategies, programmes, and projects, we highlight those with the potential to strengthen sustainability and innovativeness of Japan's capital. Finally, we address the links between being a smart and sustainable city by referring to the example of Tokyo.

2. Tokyo of the Past: Challenges and Solutions

As part of the Meiji Restoration of 1868, which set the stage for rapid modernisation following the end of the Shogunate, the city of Edo was renamed Tokyo.[4] With the relocation of the government from Kyoto, Tokyo became Japan's new and sole capital (see Matsuyama, 2022, pp. 971–975; Iwatake, 2003). The Meiji government actively promoted modern industries, leading to the establishment of several large factories along the coastline of Tokyo Bay (Sumi and Hanayama, 1985, p. 169). This aggressive reclamation of Tokyo Bay's fringe zones resulted in the creation of the Kawasaki manufacturing district and the Yokohama port, which were major projects in the region at the time (Yıldız, 2024, p. 956). Together, Tokyo, Kawasaki, and Yokohama progressively became dense hubs of heavy industries, developing rapidly and industrialising prior to their destruction by air raids during World War II (see Sumi and Hanayama, 1985, p. 169). However, these infrastructural changes, including the extensive reclamation of "waste land", i.e. the tidal and inundated areas, had a direct impact on the ecological conditions of Tokyo Bay: the water temperature, salinity levels, and aquatic habitats (see Yıldız, 2024, p. 957).

2.1 The Capital of the Japanese Economic Miracle and Its Pollution Problems

After 1945, Tokyo Bay, with its ideal natural conditions for deep harbours, played a crucial role in Japan's post-war economic boom, often referred to as "the Japanese economic miracle" (see Sumi and Hanayama, 1985, pp. 169–170). Since 1955, particularly as economic growth accelerated under the leadership of Japan's heavy industry, and with people and industries concentrating in the city, Tokyo in a short time expanded into a giant metropolis which in turn triggered the rapid reclamation of Tokyo Bay (Hotta,

2002, p. 86). Extensive land reclamation projects that utilised garbage and the dredged and mountainous soil led to the creation of 221 km^2 of new land by the 1970s, nearly 19% of the Bay's original area (see Kitazume, 2022, pp. 2–6; Sumi and Hanayama, 1985, p. 170). This transformed the coastline area into a significant industrial hub filled with numerous large-scale installations, including three steelworks, thirteen oil refineries, six petrochemical plants, and twelve other chemical facilities (Sumi and Hanayama, 1985, p. 170).

As the development of Tokyo progressed, pollution issues became evident and people's environmental perceptions began to shift (see Hotta, 2002, p. 86). In the late 1960s, under the leadership of Governor Ryokichi Minobe, the Tokyo Metropolitan Government published the White Paper 1969 on the Administration of Tokyo (1969b). This comprehensive report examined the various challenges faced by the Japanese capital. A shortened English-language edition was also released, highlighting key points from the original document (Tokyo Metropolitan Government, 1969a). The report was divided into three parts: first, on city administration; second, on bottlenecks and hurdles; and third, on meeting "the minimum standards and necessary conditions for the residents of a modern metropolis" (Tokyo Metropolitan Government, 1969a).

In the second part of the report, one finds chapter 3 on public hazards, defined "as something that adversely affects the health, wellbeing and livelihood of groups of people, and as such is a menace to society", categorised into three types: air pollution, river contamination, and excessive noise (Tokyo Metropolitan Government, 1969a, p. 75). In comparison, the Basic Law for Environmental Pollution Control (1967), in addition to these three categories (air pollution, water pollution, and noise), distinguished vibration, ground subsidence, and offensive odour as "environmental pollution" (Sumi and Hanayama, 1985, pp. 176–177).

As for the water pollution, according to the report, "Tokyo's once-famed Sumida River became so contaminated it has been dubbed the 'River of Death'". Other major rivers crucial to Tokyo's water policy, such as Tama River, Ayase River, and Meguro River, also faced significant contamination issues, primarily caused by wastewater discharged from factories (Tokyo Metropolitan Government, 1969a, p. 78). Figure 13.1 illustrates this situation using biological oxygen demand as a measure.

At that time, among the three categories of problems covered by the 1969 report (or six as under the 1967 Basic Law), the first one – air pollution – appeared to be a particularly worrying national concern. Although Japan implemented air pollution control measures before and shortly after World War II, including anti-smoke and anti-soot laws adopted by different prefectures and cities in the 1930s (Tokyo itself did so in 1935), followed by the Act on the Regulation of Soot and Smoke Emissions (1962), the situation was far from satisfactory (Sokołowski, 2022, p. 86). It did not go unnoticed by the public and the press. As newspapers reported, "Japan has won its economic battle and attained the status of a superpower in G.N.P. [Gross National

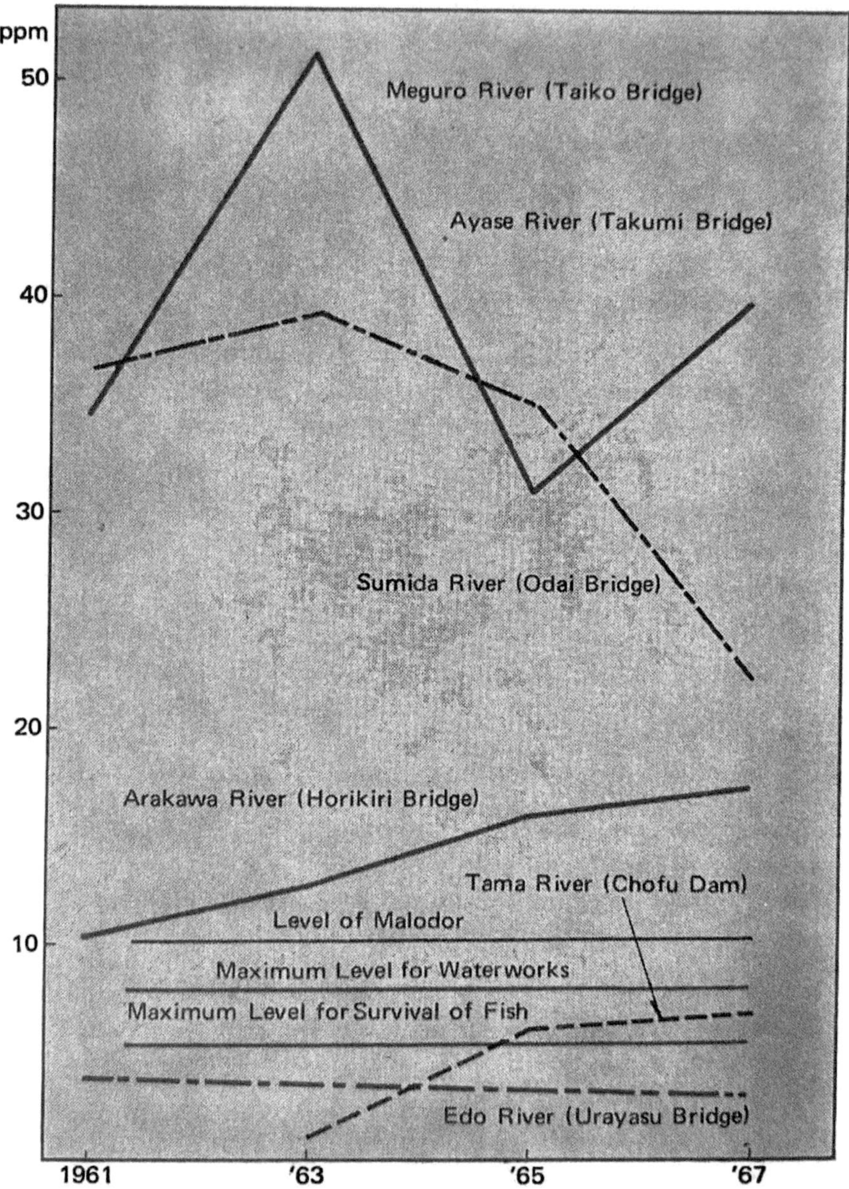

Figure 13.1 Contamination of major rivers in Tokyo

Source: Tokyo Metropolitan Government (1969a, p. 77)

Product] only to find that the slogan to which it has been so religiously dedicated means Gross National Pollution".[5]

Indeed, rising pollution was a negative consequence of Japan's rapid economic growth. While pollution existed before World War II, most Japanese people did not become seriously concerned until the 1950s and 1960s (Mori, 2008, p. 1468), when particularly harmful cases of poisoning and illness, and eventually deaths were documented (see Tamashiro *et al.*, 1984; Imai *et al.*, 1986; Kawano *et al.*, 1986). They stemmed from various sources including water pollution and food contamination (Minamata disease and Niigata Minamata disease due to methylmercury poisoning), cadmium poisoning (Itai-itai disease), and air pollution as in the case of Yokkaichi asthma (Sokołowski, 2022, p. 87). The last one was caused by sulphur dioxide (SO_2), with annual emissions exceeding 100,000 tonnes, resulting in a significant increase in the concentration of sulphur oxides (SO_x) in the polluted areas of Mie Prefecture (see Guo *et al.*, 2008; Yorifuji *et al.*, 2019).

The case of Yokkaichi and sulphur oxide emissions was also reflected by the Tokyo Metropolitan Government. As it was reported, "in front of the Metropolitan Administration Building in downtown Tokyo, the number of times the amount of sulphurous acid gas exceeded 0.3 ppm [parts per million] in 1964 was zero, while in 1967 this level was exceeded 205 times" (Tokyo Metropolitan Government, 1969a, p. 78). With the increasing consumption of heavy oil as a replacement for coal (see Figure 13.2), SO_2 emissions increased significantly during the 1960s (Imura, 2005, pp. 21–22). The 1969 report dramatically underscored the severity of the situation:

> [a]s factories have switched from coal to oil, the visible soot and cinders that turned the skies grey have been replaced by invisible, but extremely harmful, sulphurous acid gas and carbon monoxide, the volume of which has increased with the proliferation of factories, automobiles and buildings that use oil in large quantities.
>
> (Tokyo Metropolitan Government, 1969a, pp. 77–78)

Although the 1960s smog changed its structure, it was still a problem of Japan's capital. "Tokyo's smog may not be as dark as it used to be, but it certainly is more harmful", warned Tokyo Metropolitan Government (1969a, p. 78). Unfortunately, reality soon confirmed this detrimental impact on health. On 18 July 1970, a large-scale photochemical smog occurred in Tokyo, affecting over 19,000 people by 5 August, as reported to health centres (Homma, 1972, p. 6). In the following year, the first health problems associated with air pollution re-emerged on 12 May, with the total number of reported incidents exceeding 30,000, likely underestimating the actual number due to unreported cases (Homma, 1972, p. 6). The reports included health problems such as chest discomfort, eye irritation, and muscle spasms – particularly among schoolchildren, and the overall number of deaths from asthma and chronic bronchitis increased (Ziegler, 1995, p. 438).

These circumstances made the long-term slogan "give Tokyo back its sky!" even more resonant (see Tokyo Metropolitan Government, 1969a, p. 75). The pollution that attracted public attention was concentrated in specific areas, such as the heavy and chemical industrial zones located along the Pacific coastline, including Tokyo and neighbouring Yokohama (see Iwami, 2005,

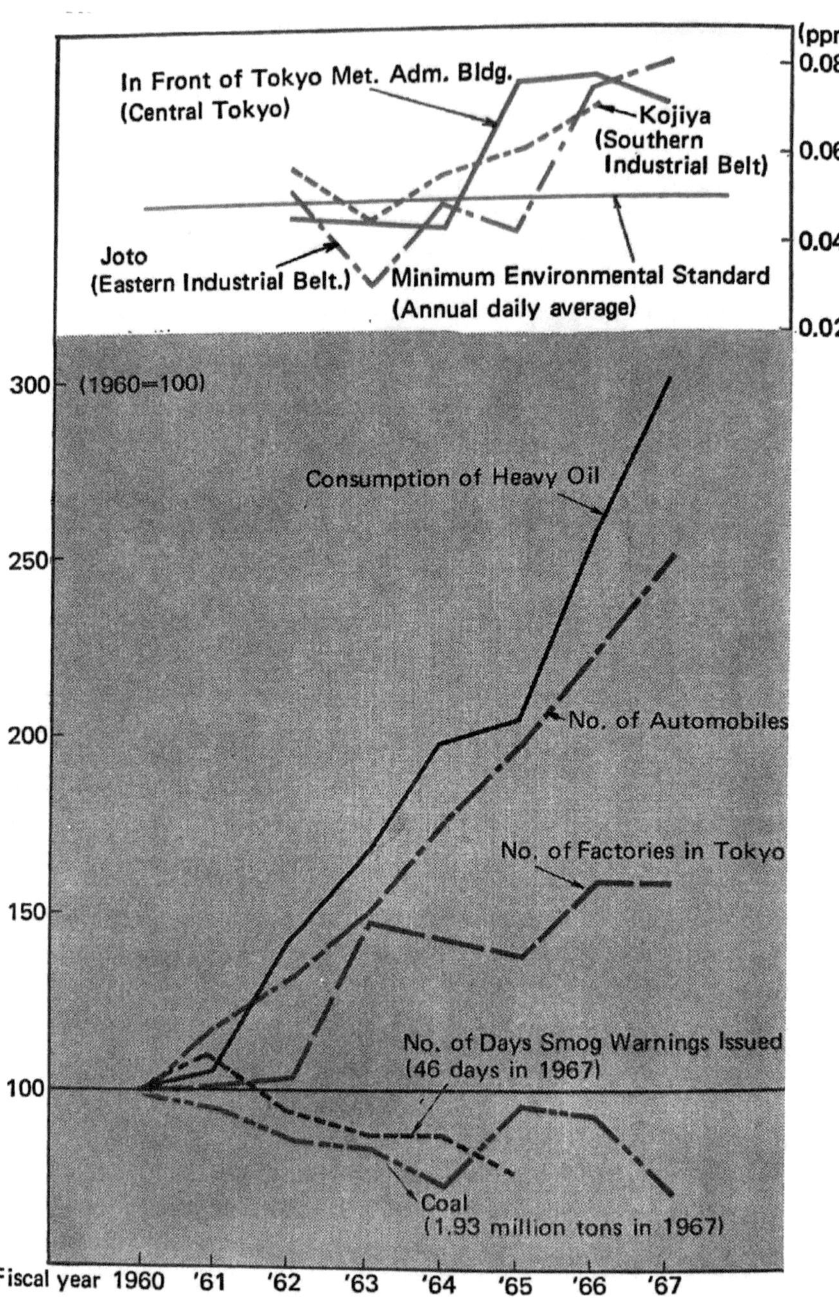

Figure 13.2 Origins and extent of the 1960s air pollution in Tokyo

Source: Tokyo Metropolitan Government (1969a, p. 76)

p. 193). The high political consciousness of residents in these areas was a driving force for anti-pollution policies, leading to civil movements that, apart from impacting national policies, also spurred local governments to take the initiative (see Iwami, 2005, pp. 192–193; Sokołowski, 2022, pp. 86–90). This happened in both the 1960s and the 1970s. For instance, in 1964, a pollution control agreement was signed between the Electric Power Development Company Limited[6] and the government of Yokohama, followed by the Tokyo Metropolitan Government signing its own pollution control agreement with the Tokyo Electric Power Company (TEPCO) in 1968 (Iwami, 2005, p. 192). To protect residents, local authorities implemented a photochemical smog warning system, issuing alerts 45 times in 1973, 41 times in 1975, and 35 times in 1984 (Shibata, 2008, p. 257). In 1974, a total emission control for SO_x was introduced in Tokyo, Yokohama, and Osaka (Iwami, 2005, p. 193). Moreover, the Tokyo Municipal Government implemented measures such as promoting fuel shift to electricity and gas and introducing district heating and cooling systems (Okata and Murayama, 2011, p. 34).

2.2 *Tokyo in the Bubble: During the Stagnation and Before the Summer Olympics*

In the 1980s, pollution problems gradually worsened in Japan. This stemmed from energy crisis, urbanisation (see Hirohara *et al.*, 1988, p. 374), and the emergence of new chemical substances not covered by existing laws (see Sprenger and Ohmura, 1997, p. 338). Significant pollution was also caused by on-road diesel vehicles, exacerbated by dense city populations, favourable diesel fuel policies from the 1970s, and lax emission controls, leading to deteriorating air quality in many cities as diesel trucks grew common in goods transport, undoing some of the central government's progress in controlling emissions from stationary sources and light-duty vehicles (Rutherford and Ortolano, 2008, p. 239). Despite these challenges, major Japanese cities tried to respond. In 1981, Tokyo, Yokohama, and Osaka adopted an emission control framework for nitrogen oxides (NO_x) similar to the 1974 framework for SO_x (see Iwami, 2005, p. 193). In 1989, the Tokyo Metropolitan Government adopted several action programmes to contribute to global environmental protection, a move that was exceptional at the international level at that time (Park, 1993, p. 40).

Moreover, in the late 1980s, Tokyo found itself at the heart of a severe crisis stemming from the collapse of the Japanese economy at the turn of the 1980s and 1990s. The enormous increases in land value associated with economic growth and the development of Japanese megacities transformed the developmental state into a construction state, where land and property development became a significant portion of economic activity, leading to a bubble and misallocation of funds that resulted in the destruction of trillions of JPY in assets, crippling Japanese financial institutions for a long time – a primary factor in the lengthy stagnation that started at the beginning of the 1990s (Sorensen, 2019, p. 32; see Iyoda, 2010, pp. 69–79). The real estate-oriented economy of the 1980s, amidst high growth, caused excessive land speculation and skyrocketing land prices, a phenomenon that started in commercial areas in Tokyo, spread to major metropolitan areas, and eventually emerged in the entire country (Tsuruta, 1999, pp. 32–33).

On the other hand, the situation improved with respect to environmental standards and a policy framework on sustainability. In the 1990s, the Japanese environmental law underwent a significant change, also influenced by international legal solutions (Nakanishi, 2016, p. 2). Moreover, climate change was also a factor in this shift. Tokyo quickly responded to this growing challenge, developing plans to make the city more sustainable, resilient, and climate-change-resistant, and urban ecology indicators were prioritised to reduce the city's environmental impact (see Lambright *et al.*, 1996, pp. 464–465).

In 1999, a soft approach to diesel pollution regulation was stopped when the Tokyo Metropolitan Government, under Governor Shintaro Ishihara, launched the "Operation No Diesel" (see Rutherford and Ortolano, 2008, p. 240; Minoura *et al.*, 2009). This campaign was a local initiative advocating stricter regulations for on-road diesel emissions, introducing an emission standard that mandated the fitting of either diesel particulate filters (DPF) or diesel oxidation catalysts (DOC) on all diesel-powered trucks, buses, and special-purpose vehicles (such as fire trucks or ambulances), in operation for more than seven years and used within city limits (Rutherford and Ortolano, 2008, p. 240). Those unable to meet the new emission standards were not allowed to drive in the Tokyo area from October 2003 onward (see Ota, 2001, p. 6, 11).

In 2000, Tokyo adopted the Metropolitan Environmental Security Ordinance to address environment-related issues (see Yamashita, 2011, p. 14). In 2002, the Tokyo CO_2 Emission Reduction Program introduced the requirements for large facilities to submit regular reports on greenhouse gas (GHG) emissions (see Asian Network of Major Cities 21, n.d.). The Program, revised in 2005, paved the way for the Tokyo Cap-and-Trade Program by collecting data from commercial and industrial sectors, obtaining their emissions reduction plans, and providing valuable experience (see Kimura, 2015, p. 16). Under the Program, more than 1,000 entities responsible for about 40% of CO_2 emissions in the industrial and business sectors of Tokyo (large-scale factories, offices, commercial facilities) submitted their plans (Okata and Murayama, 2011, p. 37).

In 2002, the Tokyo Metropolitan Government launched the Tokyo Green Building Program requiring owners of large buildings to submit Green Building Plans for improving environmental performance when reconstructing or expanding buildings; as many 1970s-era buildings were being demolished and rebuilt, there were opportunities to construct energy-saving buildings (Yoshida *et al.*, 2017, p. 142; Okata and Murayama, 2011, p. 37). In addition, in 2005, to raise environmental standards, a separate incentive was introduced for condominiums, i.e. the Tokyo Metropolitan Condominium Environmental Performance Indication Program (Okata and Murayama, 2011, p. 37). Further, in 2002, Tokyo launched a voluntary Energy Efficiency Labeling System in collaboration with major retailers, aimed at encouraging consumers to purchase energy-saving home appliances (Okata and Murayama, 2011, p. 37).[7]

In the mid-2000s, the Tokyo Climate Change Strategy was adopted (Tokyo Metropolitan Government, 2007). The primary goal of the Strategy was to reduce energy demand and facilitate a transition to a city where residents can enjoy comfortable living while consuming less energy (see Tokyo Metropolitan Government, 2007, p. 6). This involved changes in lifestyle habits, urban planning, and architectural design – to achieve this, the initial step was to encourage the introduction of energy-saving technologies and behaviours (see Tokyo Metropolitan Government, 2007, p. 6). The Strategy outlined a plan for a city emissions trading scheme, an idea that was initially proposed in 2000 and later revisited (see Tokyo Metropolitan Government, 2007, p. 14; Roppongi *et al.*, 2017, p. 521). Following consultations with a wide group of stakeholders, including representatives from industry, non-governmental organisations, and the Japanese government as observers, an agreement on the scheme was reached in 2008 (Roppongi *et al.*, 2017, p. 522). This led to the start of the emissions trading scheme in 2010 as a unique and first in the world city cap-and-trade programme specifically targeting energy-related CO_2 emissions from buildings in the commercial, industrial, and public sectors (see Nishida and Hua, 2011, p. 519; Arimura and Abe, 2021, p. 518).

However, in 2011, Japan faced a significant threat to its economy, society, energy system, and sustainability plans when a devastating earthquake and tsunami struck the Tohoku region on 11 March 2011. The Fukushima Daiichi Nuclear Power Plant, in particular, drew global attention as the condition of its reactors became extremely critical. To address the substantial decline in power capacity, both the Japanese government and electricity utilities initiated a variety of supply-side and demand-side measures, such as calls for saving energy and introduction of rolling blackouts (Sokołowski, 2022, p. 47).[8] As TEPCO, the operator of the Fukushima Daiichi Nuclear Power Plant, was Tokyo's main power supplier, the Japanese capital faced significant challenges during that period, including scheduled blackouts in the aftermath of the crisis and electricity shortages (see Aleksejeva *et al.*, 2023, pp. 77–78). In Tokyo, as in the Tohoku region, the electricity supply–demand balance was especially tight during the summer of 2011, which prompted the government to impose mandatory rationing on large customers, while households were urged to conserve electricity, particularly by reducing the operating time of air conditioners and adjusting temperature settings (see Kimura and Nishio, 2016, p. 67; Sokołowski, 2022, p. 47).

Despite an overall decrease in energy consumption, GHG emissions of the capital of Japan increased, driven by changes in the national energy mix following the Fukushima disaster (see Aleksejeva *et al.*, 2023, p. 78; Sokołowski, 2015). In contrast to the Japanese government's stance, these circumstances did not alter Tokyo's commitment to reducing GHG emissions by 25% by 2020 compared to 2000 levels – as per the 2006 pledge reaffirmed in 2013 – and further supplemented this commitment with an energy consumption target, aimed at decreasing energy consumption in Tokyo by 20% by 2020 (Aleksejeva *et al.*, 2023, p. 78). These plans were to be supported by other means such as installing renewable energy sources and investing in cogeneration (see Clarisse, 2015, p. 21).

In 2013, during the 125th Session of the International Olympic Committee (IOC) in Buenos Aires, Tokyo was announced as the host city for the 2020 Summer Olympics. "[I]n the end it was Tokyo's bid that resonated the most with the IOC membership, inviting us to 'discover tomorrow' by delivering a well-organised and safe Games that will reinforce the Olympic values while demonstrating the benefits of sport to a new generation", said the then-President of the IOC, Jacques Rogge (IOC, 2013). In the spirit of "discovering tomorrow", Tokyo promised to host environmentally friendly Olympics, emphasising sustainability in all aspects of the 2020 Games, from construction and transportation to energy and waste (see Climate Action, 2014). As the then-head of the Japanese Olympic Committee, Tsunekazu Takeda, announced:

> [i]n addition, we want the Olympic Village to become a new model for sustainable inner-city housing. Energy consumption will be minimised through the use of renewable energy sources including solar power, a seawater heat pump, use of surplus heat generated by waste treatment plants, and biogas power generation using food waste. The Olympic Village will become an urban residential "smart city pioneer model", using a wide range of Japanese sustainability technologies.
>
> (Climate Action, 2014)

Reality took an unexpected turn for both the Tokyo Olympics and the President of the Japanese Olympic Committee. In 2019, Tsunekazu Takeda resigned amidst corruption allegations (see The Japan Times, 2019). The following year, the Tokyo Games were postponed due to the COVID-19 pandemic and held in 2021 under strict health protocol and without spectators (see IOC, 2020, 2021). In this way, hailed as the "Green Olympics", the Tokyo 2020 Games achieved most of their sustainability goals primarily due to the lack of crowds (Gilson, 2023, p. 744; see Ito *et al.*, 2022).

3. Future Smart City Tokyo

In 2021, the Tokyo Metropolitan Government released Tokyo's Long-Term Strategy. Among its "strategies" being a cornerstone of Future Tokyo, one finds the realisation of Smart Tokyo via digital transformation (DX). Future Tokyo

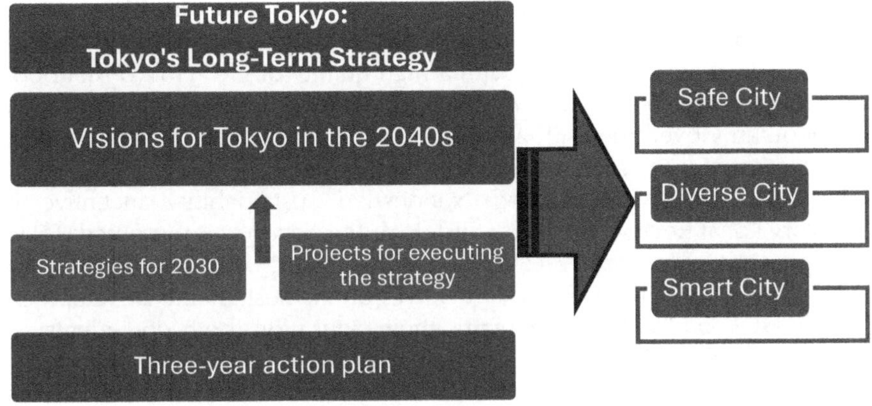

Figure 13.3 Policy correlations under the Tokyo's Long-Term Strategy
Source: Tokyo Metropolitan Government (2023, p. 11)

outlines visions for Tokyo in the 2040s. It is comprised of over 20 key concepts and 122 projects to implement them. Figure 13.3 highlights their correlations. Created in the midst of COVID-19 pandemic, the Long-Term Strategy is aimed at building a sustainable society by utilising digital technologies (Tokyo Metropolitan Government, 2021b, p. 1). Yuriko Koike, the Governor of Tokyo, highlights this process as follows:

> [t]ransitioning to carbon neutral is a global trend, and digital transformation (DX) must be advanced at an explosive speed. We will not only respond agilely to such structural changes around society, but will also have these propel sustainable growth. We will strongly promote initiatives to achieve a sustainable recovery, and transform Tokyo into the world's city of choice.
>
> (Koike, 2021)

3.1 Digital Green Transition (DGX)

In 2022, Tokyo unveiled a revised Long-Term Strategy that incorporated lessons learned from the 2020 Olympics and the COVID-19 pandemic – this Strategy was further updated in 2023 and again in 2024 (see Tokyo Metropolitan Government, n.d.b.). The latest 2024 version focuses on accelerating progress towards a sustainable future by tackling long-standing challenges and capitalising on Tokyo's unique strengths (see Tokyo Metropolitan Government, n.d.b.). To achieve this vision, Tokyo will leverage technology to address complex challenges such as reducing environmental impact while promoting economic growth and achieving harmony between nature and advanced urban functions (Tokyo Metropolitan Government, 2021b, p. 25).

In this way, Tokyo in the 2040s will be "a city that harnesses the power of digital technology to draw out its potential to realize Smart Tokyo (Tokyo's Society 5.0), providing citizens with a high quality of life" (Tokyo Metropolitan Government, n.d.a.). Smart Tokyo is a project managed by the Tokyo Metropolitan Government (2023), presented as "a convenient and comfortable city that makes use of advanced technologies".

In 2021, under the United Nations-led sustainability incentive of Voluntary Local Review, the Tokyo Sustainability Action was presented (Tokyo Metropolitan Government, 2021a). Based on the 2021 Long-Term Strategy, the 2021 action plan outlines Tokyo's initiatives on the Sustainable Development Goals (SDGs) and approach towards their realisation (both documents are complementary, while offering a bit different focus). In this way, the idea of Future Tokyo is based on three directions that can be summarised as follows: creating a sustainable city, realising swift digitalisation, and creating a new way of living (see Tokyo Metropolitan Government, 2021a, p. 9). Actions that pave the way for this direction cover, *inter alia*, advancing urban development that balances nature and convenience, promoting green growth and digital transformation, and creating a society where diverse people connect both in person and online (see Tokyo Metropolitan Government, 2021a, p. 9).

Realisation on sustainable Tokyo also involves changes with Tokyo's administration. The Tokyo Metropolitan Government (see 2021a, pp. 37–38) will implement structural reforms under the framework of Digital Transformation (DX) to digitise key procedures, reduce paper usage, eliminate seals, implement cloud computing, use cashless payments, etc., to work more efficiently and reduce bureaucracy. An idea of virtual Tokyo Metropolitan Government enhances this transition (see Tokyo Metropolitan Government, 2021a, p. 37). What complements this approach is the realisation of the TOKYO Data Highway concept, based on the full utilisation of digital technology for a sustainable and smart city, enabling connectivity to 5G networks anytime and anywhere (see Tokyo Metropolitan Government, 2021a, p. 17). Furthermore, by fusing cyber and physical spaces in a public–private partnership data platform that enables the aggregation and coordination of various data, Tokyo plans to create a digital twin environment (Tokyo Metropolitan Government, 2023, p. 18).

Moreover, within the Smart Tokyo framework, some activities on climate resilience are listed. These include improving dissemination of flood disaster information based on integrated display of rainfall and water level data with camera images linked with weather information (see Tokyo Metropolitan Government, 2021a, p. 18). In the same vein, Tokyo plans to install smart poles equipped with 5G antennas, high-speed Wi-Fi, and sensors to acquire real-time data on temperature, humidity, wind direction, and atmospheric pressure for use in combatting extreme heat, while also implementing smart water meters to monitor detailed water usage, detect leaks, and optimise waterworks operation (see Tokyo Metropolitan Government, 2021a, p. 18). On the other hand, to prevent utility poles from collapsing during earthquakes, windstorms, and floods, and to ensure a smooth response during disasters, utility poles will be gradually removed (Tokyo Metropolitan Government, 2023, p. 60). What accompanies this framework is a project of creating 3D urban model in the entire area of Tokyo (Tokyo Metropolitan Government, 2023, p. 18).

3.2 The Major Goals, Policies, and Projects

One of the key actions to reach these goals is the Strategy for Zero Emission Tokyo. It is a vision for the 2040s, when Tokyo aims to achieve zero carbon emissions by the end of that decade and rely solely on sustainable energy resources including renewables and hydrogen (see Tokyo Metropolitan Government, 2021a, p. 35). Earlier, for 2030, the policy goals considered reducing GHG and energy consumption by 50% compared to 2000, along with increasing the volume of electricity produced from renewables by half, including adding 1.3 gigawatts of power in photovoltaics (see Tokyo Metropolitan Government, 2021a, p. 35). Figures 13.4–13.6 illustrate Tokyo's 2030 pathways of progress in terms of reduction of GHG emissions and energy consumption as well as growth of renewable electricity consumption.

With respect to urban development, one should note the actions planned to make Tokyo a green city, with 130 ha of new metropolitan parks by 2030 (to reach 2,168 ha) and 107 ha of new marine parks (to reach 980 ha), using spaces along the water in areas of Asakusa, Ryogoku, Tsukuda/Etchujima,

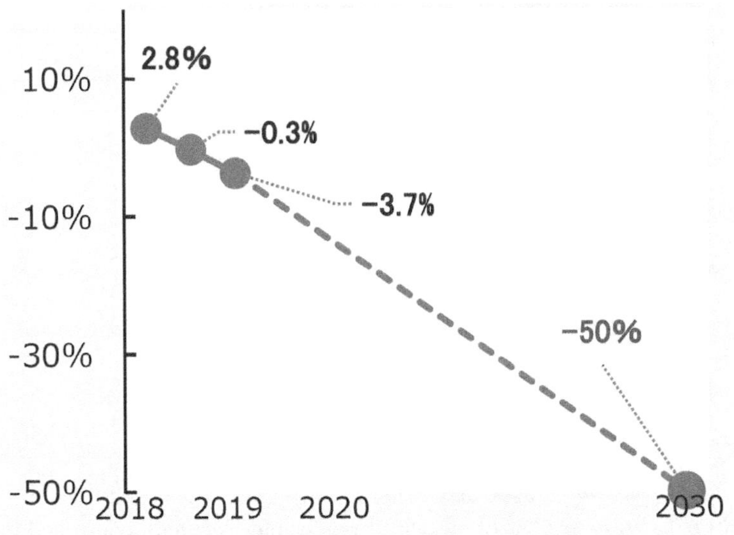

Figure 13.4 Pathway to 50% reduction of GHG emissions in Tokyo (compared to 2000)
Source: Tokyo Metropolitan Government (2023, p. 80).

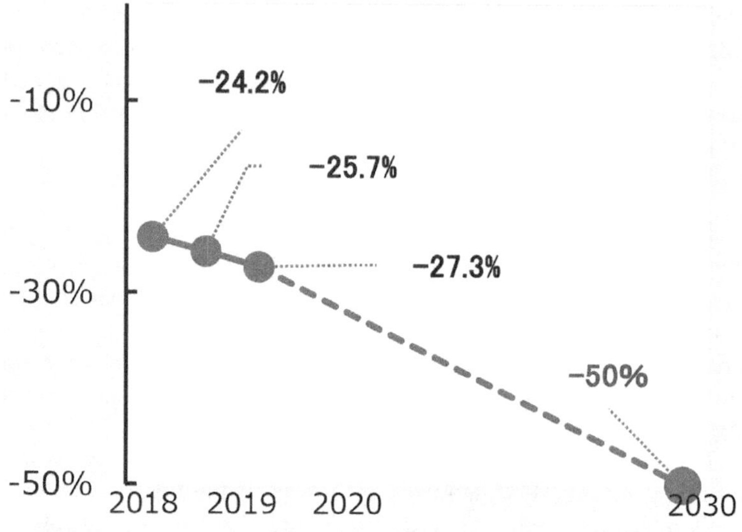

Figure 13.5 Pathway to 50% reduction of energy consumption in Tokyo (compared to 2000)

Source: Tokyo Metropolitan Government (2023, p. 80).

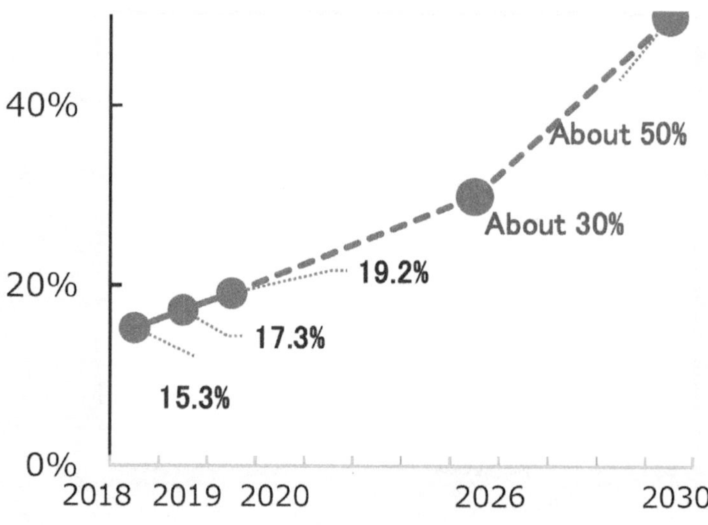

Figure 13.6 Pathway to 50% of growth of renewable electricity consumed in Tokyo (compared to 2000)

Source: Tokyo Metropolitan Government (2023, p. 80).

and Tsukiji, and completing promenades along the Sumida River by 2030 (see Tokyo Metropolitan Government, 2021a, p. 33). The main incentives for realising this action are gathered in Table 1.

An update of the 2021 Tokyo's action plan from 2023 not only reviews the progress on the goal but also highlights some related projects: Project to Realize a Tokyo Filled with Greenery; Project for Combining Community Development and Waterside Restoration; Project for Combining Community Development and Waterside Restoration; Outer Moat Cleaning Project; and Project for Providing a Stable Supply of Safe, Tasty Water and Realizing

Table 13.1 Main initiatives for making Tokyo a city filled with water and greenery.

Category	*Action*	*Stakeholders*
Urban greening and green public spaces	Create places filled with greenery	Tokyo Metropolitan Government Municipalities Private sector
	Develop parks, green spaces, etc., based on city planning	Tokyo Metropolitan Government Municipalities
	Create green spaces when private development projects are undertaken	Tokyo Metropolitan Government Private sector
	Encourage the greening of building facades and roof space	Tokyo Metropolitan Government Municipalities Private sector
Environment and nature conservation	Promote the designation of areas for the preservation of greenery	Tokyo Metropolitan Government
	Maintain an environment that supports the preservation of biodiversity	Tokyo Metropolitan Government
	Preserve and utilise productive green land, agricultural land, etc.	Tokyo Metropolitan Government Municipalities
	Create forests with less pollen	Tokyo Metropolitan Government
Waterfront greening	Promote greening at facilities along rivers	Tokyo Metropolitan Government
	Develop marine parks	Tokyo Metropolitan Government
Connected greening	Create greenery using metropolitan housing land	Tokyo Metropolitan Government Municipalities
	Form a network of greenery using existing stock	Tokyo Metropolitan Government Private sector

Source: Made by authors based on Tokyo Sustainability Action (Tokyo Metropolitan Government, 2021a, p. 34).

a Good Water Cycle (Tokyo Metropolitan Government, 2023, pp. 77–78). The last of them aims to promote the conservation of water quality in public waters while ensuring a stable supply of high-quality water for the future. This will improve the water conservation function and introduce new water treatment technology that can appropriately respond to changes in raw water quality due to climate change (Tokyo Metropolitan Government, 2023, p. 78). Although Tokyo's water quality is better than it was in the past, climate-driven challenges necessitate the implementation of new water management. Here, technologies offer solutions. Similarly, for sewage, Tokyo plans to improve the quality of treated sewage effluent by introducing advanced treatment technology to remove nitrogen and phosphorus from sewage at sewage treatment facilities (Tokyo Metropolitan Government, 2023, p. 78).

Both the 2021 Strategy and the 2021 Sustainability Action refer to the Tokyo Bay eSG Project aimed at becoming a model for a sustainable city where nature and convenience coexist, and advanced technologies are utilised in harmony with the environment (Tokyo Metropolitan Government, 2021a, p. 4). Based on four pillars – a net-zero carbon emission city abundant in water and greenery, advanced digital technology, green finance projects, and enhanced transportation networks – the Tokyo Bay eSG Project aspires to reclaim Tokyo's position as the world's most advanced city (Tokyo Metropolitan Government, 2021a, p. 16). The Project aims to employ advanced urban planning with sustainable transportation and zero-emission buildings that rely only on clean energy, such as wind power plants and floating photovoltaics, connected to a smart grid (see Tokyo Metropolitan Government, 2022, p. 7). Moreover, by leveraging cutting-edge digital technologies, Tokyo aims to transform itself into a model sustainable city, setting a benchmark for Japan and the world – this transformation targets reclaimed land in Tokyo Bay and stipulates rapid development of infrastructure to support 5G as well as the introduction of autonomous driving, Mobility as a Service (MaaS), and other innovative solutions to make the entire area a showcase (see Tokyo Metropolitan Government, 2022, p. 8).

Finally, the Tokyo Bay eSG Project incorporates an element that transcends the framework of a conventional city, or even a smart city, but in its real, physical dimension (as smart cities also utilise online resources). The Project involves the creation of a virtual city or a digital smart city. This refers to the establishment of a city operating system that will recreate the urban functions of the area in virtual reality, thereby realising the concept of a digital twin (see Tokyo Metropolitan Government, 2022, p. 8). This digital twin environment will be linked to camera images, data on the movement, and other information (see Tokyo Metropolitan Government, 2022, p. 8). This idea of the digital smart city offers a valuable platform for testing and introducing the concept of cybernetic avatars (see Ishiguro, 2021). However, addressing the complex, multifaceted, and boundary-spanning issues associated with these activities is crucial (see Shimpo, 2024). This, for instance, concerns the sustainability of such created virtual spaces and artificial intelligence (AI), as the world that has been trying to tackle climate change and environmental problems should not create new problems but should focus on solutions, such as standards for energy-efficient servers or mandates to power them with renewable energy (see Sokołowski, 2024). Frameworks covering a wide range of avatar-related problems, such as ethical, economic, environmental, legal, and social implications (E3LSI), provide promising solutions (see Shimpo, 2024).

4. Conclusion

> December 5, 1945
>
> TOKYO
>
> the closer we came to Yokohama, the plainer became the gravity of Japan's hurt. Before us, as far as we could see, lay miles of rubble. . . . There were no new buildings in sight. The skeletons of railway cars and locomotives remained untouched on the tracks. Streetcars stood where the flames had caught up with them, twisting the metal, snapping the wires over-head, and bending the supporting iron poles as if they were made of wax. Gutted buses and automobiles lay abandoned by the roadside. This was all a man-made desert, ugly and desolate and hazy in the dust that rose from the crushed brick and mortar.
>
> (Gayn, 1981, pp. 1–2)

The post-war outlook of Japan, as in the cited 1945 report from the road to destroyed Tokyo, was not optimistic. The landscape of destruction inspired more fear than hope. Fortunately, the following years of post-war reconstruction brought a different scenario. Japan not only managed to rise from the ashes but also grew in an incredibly dynamic way, surprising the world. This happened due to a combination of exceptionally favourable circumstances and the hard and intensive work of the Japanese people, their persistence, pride, and choices.

Nevertheless, development did not take place without costs. Along with economic progress came burdens on the natural environment resulting in its pollution. This growing problem of the 1960s and 1970s contributed to changes, including regulatory ones. The increased awareness of environmental problems within Japanese society became a significant compass for future directions. This issue permanently entered the political agenda, gradually shifting its focus towards sustainability and climate change.

The case of Tokyo, as history has demonstrated, highlights these processes well. This mirror-like feature of Tokyo, accumulating all that Japan represents, extends far beyond its borders. This characteristic of being a benchmark is the remarkable strength of Japan's capital. Its influence reaches widely across Asia and globally, spanning various socio-economic domains, including energy and climate or green and digital transformation. Therefore, to implement changes in these areas, it is worth examining the case of Tokyo as an insightful lesson offered by a smart global metropolis aiming for sustainability, a city that always wants to be a day ahead – and this ambition is not solely due to its time zone.

However, this feature is not a spontaneous occurrence. It is the culmination of numerous factors, including successes, failures, wise, and less-wise decisions, as well as the ambition and ingenuity of Tokyo's residents and leaders, which have made Tokyo a global point of reference.

This also encompasses initiatives related to sustainability and smart city policies. In this context, Tokyo is a city that utilises new technologies to enhance services for residents, improves management in various fields like planning or administration, and addresses challenges to create new value, while striving to achieve sustainability and realise the vision of Society 5.0, ahead of its peers. Tokyo not just meets the Japanese criteria for being a *smart city*; it also actually sets those standards in many places. However, this race has no single leader, and in the era of the AI revolution, each day brings new solutions and challenges. Hence, it is important to maintain a brisk pace. So far, Tokyo is doing quite well.

Acknowledgements

This research was supported by the JST Moonshot R&D project, JPMJMS2215.

Notes

1 Approximately 700 billion USD.
2 Approximately 102 billion USD.
3 The Society 5.0 (the super smart society) is the new society created by transformations led by scientific and technological innovation, after hunter–gatherer society – Society 1.0, agricultural society – Society 2.0, industrial society – Society 3.0, and information society – Society 4.0 (see Shimpo, 2018, p. 49).
4 In Japanese: 東京 translates to "eastern capital".
5 https://time.com/archive/6838804/japan-a-blue-sky-for-tokyo/
6 電源開発株式会社 [*dengen kaihatsu kabushiki-gaisha*] is a direct English translation of the company's Japanese name. The company is more commonly known by its brand name: J-POWER.
7 In 2005, the Energy Efficiency Labeling System for Home Appliances became mandatory (Okata and Murayama, 2011, p. 37).
8 The rolling blackouts introduced in March 2011 were the first scheduled blackouts in Japan since the immediate aftermath of World War II and were eventually discontinued in April 2011 (Hayashi and Hughes, 2013, p. 89).

References

Act on the Regulation of the Emission of Soot and Smoke (Act No. 146 of 1962) In Japanese: 煤煙排出規制法 [*baien haishutsu kisei-hō*].

Aleksejeva, J., Voulgaris, G., Long, Y. and Gasparatos, A. (2023) "Co-Evolution of Energy and Climate Change Mitigation Policies in Japan and Tokyo", in P.J. Marcotullio, J. Sperling and A.L. Pierce (eds) *Urban Energy And Climate: Prospects for a Sustainable Transition*. Singapore: World Scientific (World Scientific Series in Current Energy Issues), pp. 63–93.

Arimura, T.H. and Abe, T. (2021) "The Impact of the Tokyo Emissions Trading Scheme on Office Buildings: What Factor Contributed to the Emission Reduction?', *Environmental Economics and Policy Studies*, 23(3), pp. 517–533.
Asian Network of Major Cities 21 (n.d.) "Best Practices of ANMC21 Member Cities". Available at: https://www.sp.metro.tokyo.lg.jp/seisakukikaku/gaimubu/anmc21/anmc21org/english/bestpractice/Tokyo1.html (Accessed: 29 July 2024).
Basic Law for Environmental Pollution Control (1967) No. 132. In Japanese: 公害対策基本法 [*kōgai taisaku kihon-hō*].
Cabinet Office (2021) "Smart City Guidebook". Available at: https://www8.cao.go.jp/cstp/society5_0/smartcity/01_scguide_eng_1.pdf (Accessed: 22 July 2024).
Clarisse, P. (2015) "Tokyo Smart City Development in Perspective of 2020 Olympics". Tokyo. Available at: https://www.eu-japan.eu/sites/default/files/publications/docs/smart2020tokyo_final.pdf (Accessed: 27 July 2024).
Climate Action (2014) "The Green Games: Tokyo 2020" [Interview With the President of the Japanese Olympic Committee, Tsunekazu Takeda]. Available at: https://www.climateaction.org/climate-leader-interviews/the_green_games_tokyo_2020 (Accessed: 30 July 2024).
Gayn, M. (1981) *Japan Diary*. Rutland, VT – Tokyo: Charles E. Tuttle.
Gilson, J. (2023) "From Kyoto to Glasgow: Is Japan a Climate Leader?", *The Pacific Review*, 36(4), pp. 723–754.
Guo, P., Yokoyama, K., Suenaga, H.M. and Kida, H. (2008) "Mortality and Life Expectancy of Yokkaichi Asthma Patients, Japan: Late Effects of Air Pollution in 1960–70s", *Environmental Health*, 7(1), pp. 1–10.
Hayashi, M. and Hughes, L. (2013) "The Policy Responses to the Fukushima Nuclear Accident and Their Effect on Japanese Energy Security", *Energy Policy*, 59, pp. 86–101.
Hirohara, M., Alden, J.D. and Cassim, M. (1988) "The Impact of Recent Urbanisation on Inner City Development in Japan", *The Town Planning Review*, pp. 365–381. https://doi.org/10.3828/tpr.59.4.l063868061009585
Homma, Y. (1972) *Environmental Pollution in Japan: The Case of Tokyo*. Berkeley, CA: Lawrence Berkeley National Laboratory.
Hotta, K. (2002) "Tokyo Bay Reformation", in J. Chen et al. (eds) *Engineered Coasts*. Dordrecht: Springer, pp. 85–102.
Imai, M., Yoshida, K. and Kitabatake, M. (1986) "Mortality From Asthma and Chronic Bronchitis Associated With Changes in Sulfur Oxides Air Pollution", *Archives of Environmental Health: An International Journal*, 41(1), pp. 29–35.
Imura, H. (2005) "Japan's Environmental Policy: Past and Future", in H. Imura and M. Schreurs (eds) *Environmental Policy in Japan*. Cheltenham: Edward Elgar Publishing, pp. 15–48.
IOC (2013) "IOC Selects Tokyo as Host of 2020 Summer Olympic Games". Available at: https://olympics.com/en/news/ioc-selects-tokyo-as-host-of-2020-summer-olympic-games (Accessed: 30 July 2024).
IOC (2020) "IOC and Tokyo 2020 Joint Statement: Framework for Preparation of the Olympic and Paralympic Games Tokyo 2020 Following Their Postponement to 2021". Available at: https://olympics.com/ioc/news/ioc-and-tokyo-2020-joint-statement-framework-for-preparation-of-the-olympic-and-paralympic-games-tokyo-2020-following-their-postponement-to-2021 (Accessed: 30 July 2024).
IOC (2021) "Joint Statement on Spectator Capacities at the Olympic Games Tokyo 2020". Available at: https://olympics.com/en/news/joint-statement-on-spectator-capacities-at-the-olympic-games-tokyo-20-tokyo-2020 (Accessed: 30 July 2024).
Ishiguro, H. (2021) "The Realisation of an Avatar-Symbiotic Society Where Everyone Can Perform Active Roles Without Constraint", *Advanced Robotics*, 35(11), pp. 650–656.
Ito, E., Higham, J. and Cheer, J.M. (2022) "Carbon Emission Reduction and the Tokyo 2020 Olympics", *Annals of Tourism Research Empirical Insights*, 3(2), p. 100056.

Iwami, T. (2005) "The 'Advantage of Latecomer' in Abating Air-Pollution: The East Asian Experience", *International Journal of Social Economics*, 32(3), pp. 184–202.

Iwatake, M. (2003) "From a Shogunal City to a Life City: Tokyo Between Two Fin-De-Siècles", in N. Fieve and P. Waley (eds) *Japanese Capitals in Historical Perspective: Place, Power and Memory in Kyoto, Edo and Tokyo*. London: Routledge, pp. 233–256.

Iyoda, M. (2010) *Postwar Japanese Economy: Lessons of Economic Growth and the Bubble Economy*. New York, NY: Springer.

The Japan Times (2019) "Tsunekazu Takeda to Leave JOC and IOC Posts Amid Scandal". Available at: https://www.japantimes.co.jp/sports/2019/03/19/olympics/tsunekazu-takeda-leave-joc-ioc-posts-amid-scandal-tokyo-governor-confirms/ (Accessed: 30 July 2024).

Kawano, S., Nakagawa, H., Okumura, Y. and Tsujikawa, K. (1986) "A Mortality Study of Patients With Itai-Itai Disease", *Environmental Research*, 40(1), pp. 98–102.

Kimura, M. (2015) "The Tokyo Cap-and-Trade Program: A City-Level Initiative Toward a Low Carbon Society", in C. Serre et al. (eds) *Emissions Trading Worldwide: International Carbon Action Partnership (ICAP) Status Report 2015*. Berlin: ICAP, pp. 16–17. Available at: https://icapcarbonaction.com/system/files/document/icap_report_2015_02_10_online_version.pdf (Accessed: 29 July 2024).

Kimura, O. and Nishio, K.-I. (2016) "Responding to Electricity Shortfalls: Electricity-Saving Activities of Households and Firms in Japan After Fukushima", *Economics of Energy & Environmental Policy*, 5(1), pp. 51–72.

Kitazume, M. (2022) "Sustainable Land Reclamation in Coastal Area", *Revue Française de Géotechnique* [French Journal of Geotechnical Engineering], 170(2), pp. 1–15.

Koike, Y. (2021) "Policy Speech by the Governor of Tokyo, Koike Yuriko, at the Fourth Regular Session of the Tokyo Metropolitan Assembly, 2021–3. Realizing a Sustainable Recovery by Leveraging Decarbonization and Digitalization", Tokyo Metropolitan Government. Available at: https://www.metro.tokyo.lg.jp/english/governor/speeches/2021/1130/03.html (Accessed: 29 July 2024).

Lambright, W.H., Chjangnon, S.A. and Harvey, L.D.D. (1996) "Urban Reactions to the Global Warming Issue: Agenda Setting in Toronto and Chicago", *Climatic Change*, 34(3–4), pp. 463–478.

Matsuyama, M. (2022) "Edo-Tokyo and the Meiji Revolution", *Journal of Urban History*, 48(5), pp. 966–987.

Minoura, H., Takahashi, K., Chow, J.C. and Watson, J.G. (2009) "Atmosphere Environment Improvement in Tokyo by Vehicle Exhaust Purification", in C.A. Brebbia and V. Popov (eds) *Air Pollution XVII*. Southampton: WIT Press, pp. 129–140.

Mori, M. (2008) "Environmental Pollution and Bio-Politics: The Epistemological Constitution in Japan's 1960s", *Geoforum*, 39(3), pp. 1466–1479.

Nakanishi, Y. (2016) "Introduction: The Impact of the International and European Union Environmental Law on Japanese Basic Environmental Law", in Y. Nakanishi (ed) *Contemporary Issues in Environmental Law: The EU and Japan*. Tokyo: Springer, pp. 1–15.

Nishida, Y. and Hua, Y. (2011) "Motivating Stakeholders to Deliver Change: Tokyo's Cap-and-Trade Program", *Building Research & Information*, 39(5), pp. 518–533.

Okata, J. and Murayama, A. (2011) "Tokyo's Urban Growth, Urban Form and Sustainability", in A. Sorensen and J. Okata (eds) *Megacities: Urban Form, Governance, and Sustainability*. Tokyo: Springer, pp. 15–41.

Ota, K. (2001) "Requirements for Increasing Ownership of Clean Vehicles and Subjects: A Case Study of CNG Trucks", *Regular Meeting for Briefing Research Report*, 367. Tokyo: IEEJ. Available at: https://eneken.ieej.or.jp/data/en/data/old/pdf/otarp.pdf (Accessed: 27 July 2024).

Park, J. (1993) *Development of the Global Warming Agenda in Japan*. PhD Thesis. Massachusetts Institute of Technology. Available at: https://dspace.mit.edu/bitstream/handle/1721.1/69313/29456659-MIT.pdf?sequence=2 (Accessed: 27 July 2024).

Roppongi, H., Suwa, A. and De Oliveira, J.A.P. (2017) "Innovating in Sub-National Climate Policy: The Mandatory Emissions Reduction Scheme in Tokyo", *Climate Policy*, 17(4), pp. 516–532.

Rutherford, D. and Ortolano, L. (2008) "Air Quality Impacts of Tokyo's On-Road Diesel Emission Regulations", *Transportation Research Part D: Transport and Environment*, 13(4), pp. 239–254.

Shibata, T. (2008) "Social Costs of Traffic Congestion in Developing Metropolises", in *Urban Justice & Sustainability Conference*, 22–25 August 2007, Vancouver. Available at: https://repository.tku.ac.jp/dspace/bitstream/11150/645/1/keizai257-07.pdf (Accessed: 28 July 2024).

Shimpo, F. (2024) "What Are E3LSI Issues in Cybernetic Avatars", in *Ethical, Legal and Social Issues for Symbiotic Society With AI and Robots: Proceedings of the Ninth International Conference on Robot Ethics and Standards. 9th ICRES*, 29–31 July 2024, Yokohama: CLAWAR, pp. 134–137.

Sokołowski, M.M. (2015) "Priorities of Energy Policy of Japan Under Abenomics", in M. Sitek and M. Łęski (eds) *Opportunities for Cooperation Between Europe and Asia*. Józefów: WSGE, pp. 227–240.

Sokołowski, M.M. (2022) *Energy Transition of the Electricity Sectors in the European Union and Japan: Regulatory Models and Legislative Solutions*. Cham: Palgrave Macmillan.

Sokołowski, M.M. (2024) "Cybernetic Avatars, Robots, and Sustainability", in *Ethical, Legal and Social Issues for Symbiotic Society With AI and Robots: Proceedings of the Ninth International Conference on Robot Ethics and Standards, 9th ICRES*, 29–31 July 2024, Yokohama: CLAWAR, pp. 138–144.

Sorensen, A. (2019) "Tokaido Megalopolis: Lessons From a Shrinking Mega-Conurbation", *International Planning Studies*, 24(1), pp. 23–39.

Sprenger, R.-U. and Ohmura, T. (1997) "Environmental Problems", in R. Osterkamp and K. Takahashi (eds) *A Comparative Analysis of Japanese and German Economic Success*. Tokyo: Springer.

Sumi, K. and Hanayama, K. (1985) "Existing Institutional Arrangements and Implications for Management of Tokyo Bay", *Natural Resources Journal*, 25(1), pp. 167–194.

Tamashiro, H., Akagi, H., Arakaki, M., Futatsuka, M. and Roht, L.H. (1984) "Causes of Death in Minamata Disease: Analysis of Death Certificates", *International Archives of Occupational and Environmental Health*, 54, pp. 135–146.

Tokyo Metropolitan Government (1969a) *Sizing Up Tokyo: A Report on Tokyo Under the Administration of Governor Ryokichi Minobe*. Tokyo: Dai-Nippon.

Tokyo Metropolitan Government (1969b) 東京を考える – 都政白書 [*Thinking About Tokyo – Metropolitan Government White Paper*]. Tokyo: Tokyo Metropolitan Government Planning and Coordination Bureau.

Tokyo Metropolitan Government (2007) 東京都気候変動対策方針「カーボンマイナス東京10 年プロジェクト」基本方針 [*Tokyo Climate Change Strategy "Carbon Minus Tokyo 10 Year Project" Basic Policy*]. Available at: https://www.kankyo.metro.tokyo.lg.jp/documents/d/kankyo/climate-climate_change-files-honnbunn-all (Accessed: 26 July 2024).

Tokyo Metropolitan Government (2021a) "Tokyo Sustainability Action". Available at: https://www.metro.tokyo.lg.jp/english/about/sustainable/documents/tokyo_sustainability_action.pdf (Accessed: 27 July 2024).

Tokyo Metropolitan Government (2021b) "「未来の東京」戦略" [*Future Tokyo: Strategy*]. Available at: https://www.sp.metro.tokyo.lg.jp/seisakukikaku/mirainotokyo-senryaku/html5.html#page=1 (Accessed: 26 July 2024).

Tokyo Metropolitan Government (2022) "Tokyo Bay eSG Project Creating Japan's Future From Tokyo Bay (Version 1.0)". Available at: https://www.tokyobayesg.metro.tokyo.lg.jp/esgproject/pdf/Tokyo_Bay_eSG_Project_ver1.0_EN.pdf (Accessed: 30 July 2024).

Tokyo Metropolitan Government (2023) "Tokyo Sustainability Action". Available at: https://www.metro.tokyo.lg.jp/english/about/sustainable/documents/tokyo_sustainability_action2023.pdf (Accessed: 27 July 2024).

Tokyo Metropolitan Government (n.d.a.) "Future Tokyo: Tokyo's Long-Term Strategy, Tokyo Metropolitan Government". Available at: https://www.metro.tokyo.lg.jp/english/about/policies/policies01.html (Accessed: 29 July 2024).

Tokyo Metropolitan Government (n.d.b.) "Future Tokyo: Tokyo's Long-Term Strategy Version Up 2022, Tokyo Metropolitan Government". Available at: https://www.metro.tokyo.lg.jp/english/about/futuretokyo/versionup.html (Accessed: 29 July 2024).

Tsuruta, H. (1999) "The Bubble Economy and Financial Crisis in Japan", *International Journal of Political Economy*, 29(1), pp. 26–48.

Yamashita, N. (2011) "The Role of Tokyo in Japanese Climate Policy: From the Viewpoint of Multi-Level Governance", in *2010 Berlin Conference on the Human Dimensions of Global Environmental Change: "Social Dimensions of Environmental Change and Governance"*, 8–10 October 2010, Berlin: Berlin Free University. Available at: https://refubium.fu-berlin.de/bitstream/handle/fub188/17811/Yamashita-Role_of_Tokyo_in_Japanese_Climate_Policy_-From_the_Viewpoint_of_Multi-level_Governance-419.pdf?sequence=1&isAllowed=y (Accessed: 27 July 2024).

Yıldız, Ş. (2024) "Reclaimed Ecotones in the Climate Change Era: A Long-Durée Framing of Urban Expansion in Mumbai, Amsterdam, New York, and Tokyo", *Environment and Planning E: Nature and Space*, 7(2), pp. 950–974.

Yorifuji, T., Kashima, S., Suryadhi, M.A.H. and Abudureyimu, K. (2019) "Acute Exposure to Sulfur Dioxide and Mortality: Historical Data From Yokkaichi, Japan", *Archives of Environmental & Occupational Health*, 74(5), pp. 271–278.

Yoshida, J., Onishi, J. and Shimizu, C. (2017) "Energy Efficiency and Green Building Markets in Japan", in N.E. Coulson, Y. Wang, and C.A. Lipscomb (eds) *Energy Efficiency and the Future of Real Estate*. New York, NY: Palgrave Macmillan, pp. 139–159.

Ziegler, J. (1995) "Rays of Hope in the Land of the Rising Sun", *Environmental Health Perspectives*, 103(5), pp. 436–440.

Index

For Product Safety Concerns and Information please contact our EU representative GPSR@taylorandfrancis.com Taylor & Francis Verlag GmbH, Kaufingerstraße 24, 80331 München, Germany

Batch number: 10397794

Printed by Printforce, the Netherlands